Nichols, Naomi M

Food drying at home

FOOD DRYING AT HOME

Naomi M. Nichols

FOOD DRYING AT HOME

VNR
VAN NOSTRAND REINHOLD COMPANY
NEW YORK CINCINNATI TORONTO LONDON MELBOURNE

Library of Congress Catalog Card Number 77-1135
ISBN 0-442-26028-8 cloth
ISBN 0-442-26029-6 paper

Printed in Great Britain

Published in 1978 by Van Nostrand Reinhold Company
A division of Litton Educational Publishing, Inc.
135 West 50th Street, New York, NY 10020, U.S.A.

16 15 14 13 12 11 10 9 8 7 6 5 4 3 2 1

Library of Congress Cataloging in Publication Data

Nichols, Naomi M
Food drying at home.

1. Food–Drying. 2. Cookery (Dried foods)
I. Title.
TX609.N49 1977 641.4'4 77-1135
ISBN 0-442-26028-8
ISBN 0-442-26029-6 pbk

Contents

1 Why People Dry Food

When the astronauts went into space, they had a nation's resources at their disposal. Intensive research was done on their food, yet there was never any argument about what form it should take—it was always dried.

When such great distances are involved, the problems of developing the type of canned or frozen foods that would fulfil the specialized requirements of space travel would be tremendous. Using dried food eliminates the problems of spoilage due to bacterial contamination. It takes up less space and weighs less. In addition, it retains a high degree of flavour and nutrition. And, where so many crucial last-minute arrangements and decisions must be made, dried food can be prepared and stored well in advance.

But dried food is not just a spin off from the space industry. It is an ancient practice which has been with us as long as there has been sunshine to dry food and people to eat it. Even in countries like Great Britain with minimum sunshine, people dried some foods. We have heard country people from areas such as Dorset and Yorkshire tell how their mothers would string apple rings across the kitchen or over the stove to dry. Other fruits, as well as some vegetables and of course herbs, were dried in the family's airing cupboard, then stored with the meat already hanging in the attic.

In the early days of the American west, dried apples and other fruits were popular because fresh fruit was often unavailable in the sparsely settled frontier towns. When Lewis and Clark started their historic journey across the North American continent to the state of Oregon, they also had the nation's resources at their disposal. According to their journals dried soup mix was an important part of their supplies. They chose dried food because it would transport easily and needed a minimum of care to keep it from spoiling.

In Vermont winter apples were stored in basements and lasted fairly well. But to preserve apples really well, the pioneer natives would string them on sturdy thread and hang them around the fireplace, a practice that was probably brought over from Great Britain. The resulting fragrance puts modern room deodorizers to shame. This tradition is still followed, and the dried fruit makes many a pie and apple tart long after the fresh apples have been eaten.

Until recently most drying was done with the sun as the only source of heat. Occasionally fires were built under food either to smoke it or to hasten the drying process. But in semi-tropical countries like Mexico, where fresh food spoils in a day, peppers and meat have long been preserved by being hung outside the house from the roof. However in countries where days of uninterrupted sunshine are few and far between, this simply is not practical.

Certainly if there was no other way to put food aside for the winter, man made as good use of the sun's rays as he could. Farmers, of course, have always depended on the sun to dry their hay. In southern California we saw acres of apricots, peaches and prunes spread out on trays in the sun to dry. But this is in a state where the sun is a reliable and important feature of the economy.

Sun drying for the homeowner, especially the city dweller, is usually inefficient and time-consuming. It can take from one to four weeks to dry a batch of food. Trays must be carried in every night or covered, because of moisture. Dust and insects are always a problem.

But in spite of these difficulties, this oldest form of food preservation was well known and much used by our grandparents' generation. Now many of today's young people, returning to some of the old ways, are also turning to drying food, and are bringing new equipment and technology to this ancient process.

To some, this may seem like turning back the clock and giving up the advantages that modern technology has showered on the lucky citizens of the industrialized nations. It could even seem to mean returning to an era of inconvenience and a lowering of living standards. Why, with frozen and tinned foods so readily obtainable, and with the new freeze-dried products now flooding the shelves of our food shops, should we bother to find out about home food drying?

One reason may be found in this report from a study made by Georgetown University:

> Some chemical compounds used to preserve beverages and canned and frozen foods are found to be harmful to human cells. The laboratory study showed that some food additives act just as strongly against human and animal cells as they do against bacteria.

In an age when more and more synthetic substances are being poured into our air, our food and our water, the more safeguards we can personally build into our lives, the better off we will be. At least you can be sure that whatever foods you home-dry will be free from the harmful compounds found in many commercially prepared foods, as you need no preservatives to keep your dried food safe and ready for use.

Modern self-contained home dryers can now be built or purchased, and they make drying easy and pleasant. Most food will dry in twenty-four to forty-eight hours losing two-thirds to four-fifths of their original bulk. No energy is required for storage as it is with frozen food. Because dried food is so light and compact it transports easily, a great advantage when moving house.

Modern families who take on the high initial cost of a freezer, with the consequent increase in their household electricity bill, do so because they want the convenience and economy of frozen food. There is no doubt that they do gain added convenience and sometimes can save money by using their freezer. We have used both our dryer and a freezer and in Chapter 10 we compare the two methods of food preservation and give some suggestions on how to use them together to greater advantage.

When freezers first came out there was an initial period of uncertainty as consumers investigated whether or not this new appliance was suitable for them. Certainly the expense should make one weigh the advantages against the cost. So it should be with food drying, although the initial outlay for equipment is of course much less.

In almost every case dried food will give the average household advantages comparable to modern convenience foods, the only difference being that some dried foods must be reconstituted (by soaking, or simmering until soft) before cooking. But regardless of a power failure which might defrost your freezer, your dried food will remain intact, capable of helping your family through an emergency.

Commercial methods like spray drying, vacuum drying or freeze drying are complex and very expensive. They require immense outlays of money for equipment, none of which is necessary for home purposes. So is home-dried food as good to eat as commercially dried food? Taste for yourself and decide—the flavour will convince you that it is. Just as home-canned or home-bottled food tastes better than shop-bought produce so properly home-dried food tastes better than the commercial product.

A home food dryer can be built easily and at very low cost, and is easy to operate. The results are a revelation in convenience, taste and economy. Dried fruits like apricots, peaches, dates and figs have always been favourites, and they make a very acceptable substitute for sweets. The flavour of the fruit is concentrated, as much as five times, all in one chewy morsel.

Dried fruit, when reconstituted, should turn out almost exactly like canned fruit, but with several advantages. It needs no addition of sugar to preserve it. This not only saves money, but also gives the housewife greater flexibility if any members of the family are on a sugar-free diet. You remove from the packet only the amount you need. The unused portion is returned to its storage tin, to keep until needed again. This means no more half-empty tins or jars to spoil in the refrigerator, a problem familiar to every housewife. Everything you store is used, so you not only save money, but you also get more from the produce you dry and put away.

The safe preservation of vegetables is also a revelation to the woman who bottles her own foods. You have no worry about botulism with dried vegetables. And you can forget about the substantial investment in jars, lids and, in the case of fruit, sugar, that bottling requires. And of course far less storage space is needed.

A whole stew, or meal-size soup mix, can be dried and packaged so that all you need do is empty a single bag into a pot of boiling water, then add your meat or fish and, if you wish, one or two fresh vegetables. The flavour of dried vegetables compares veıy favourably with fresh.

Think what this means. The storage life of most freezer food is limited, usually from three months to a year. A slow deterioration sets in almost as soon as the food is placed in the freezer and continues until the food is nearly tasteless. Even tinned food will not keep indefinitely, especially if stored in warm areas.

But properly packed and stored dried food will keep for months and years. I have pears I dried nearly four years ago, and they are still as sweet and fragrant as when I first put them away.

In food drying water is the enemy because it not only leaches out flavour but provides an essential factor for the growth of bacteria and other agents that spoil food. Drying removes almost all of the moisture, making dried foods far safer than either tinned or frozen.

Drying means a housewife is not limited by how much freezer space she has, or by how many jars and lids she can get her hands on during the peak season. She can take advantage of fresh produce when it is at its lowest price and also at the peak of quality. Fresh fruit and vegetables usually cost as little as one-fifth of the winter prices. So the housewife can use what would be expensive food in the off season, knowing it is costing her much less than if she had bought it in the shop that day.

The question of what to dry can depend on such factors as family food tastes, what fresh foods are available to you during the summer (which will depend in large measure on where you live), and on whether you have facilities for such alternative methods of food storage as cellars, basements, or outdoor pit storage, such as clamps. Most modern urban families have no access at all to such facilities, and therefore home food drying provides them with a welcome and convenient way to store food.

Even those families who have the above-mentioned storage facilities may find they would benefit from drying foods that store well, because of the longer life, easy reconstitution and minimal storage space needed. Potatoes, carrots, and onions, for example, all dry well and can be stored fresh with equal success.

When we lived on a large farm in Oregon, we found drying more convenient. I remember one winter a friendly neighbour gave me a couple of hundred pounds of freshly dug carrots. I promptly took them home and, following his instructions, quickly replanted them in our back garden.

Yes, we had fresh carrots all winter, but at what a cost! Whenever I fancied carrots I had to slog out into the snow and mud, then spend an hour cleaning up the carrots as well as myself. It just wasn't worth spending all that time and effort for a bunch of carrots.

Most modern urban families would find such a procedure time-consuming and unpleasant. With the choice of dryers to be found in this book, drying is now feasible for everyone. The list on page 13 shows foods that can be dried successfully. Those marked with an asterisk are best, both for reconstitution and storage life. Some authorities disagree with some of the foods we have marked in this way. However, we have included only those foods we have either tried ourselves or tasted in the kitchens of friends. As an example, we note that strawberries are often disparaged as unsuitable for food drying because of their excessive juiciness. But we have reports from families who dry these luscious berries every year—they live in the heart of strawberry country—and they say they are delicious and keep their flavour well.

The key to successful and enjoyable drying is: first learn the principles, then experiment. Like anything worth doing it is a process that requires understanding, the proper equipment and time, but the results are well worth the effort. The rewards are less money spent for food, better-tasting and more healthful meals, and a real feeling of self-satisfaction.

Foods to Dry at Home

Those marked with an asterisk are most suitable for drying.

Fruit

Apples*
Apricots*
Bananas*
Berries
Cherries
Coconut
Dates
Figs*
Grapes*
Guavas
Nectarines
Papaya*
Peaches*
Pears*
Plums*
Prunes*
Pineapple
Rhubarb (stalks only)
Lemon and Orange rind (only if unsprayed fruit is used.)

Vegetables

Beans (green or snap; runner beans in Great Britain)*
Beetroot (fresh)
Beet greens
Broccoli
Cabbage*
Celery
Celery leaves*
Carrots*
Corn (maize)*
Courgettes* (small zucchini squash)
Cucumbers*
Garlic (usually purchased already dried, however)
Marrows (large zucchini)*
Mushrooms*
Onions*
Parsley*
Parsnips
Peas and all other legumes*
Peppers and pimentos (including chillies)*
Potatoes*
Pumpkin (and other squash)
Spinach (and other greens such as kale)
Squash, winter
Rutabagas (a member of the turnip family)
Turnips

Tomatoes*
Zucchini (see above)*

Nuts
All nuts

Meat

Fish

Fowl and Game

A simple rule is: dry everything that tastes good to you, with the exception of lettuce and citrus fruits. (After we wrote this, we met some Persian friends who said they dry whole lemons and use them for cooking.) For additional information on the nutritive value of dried food compared with fresh, we refer you to the following publications, as recommended by the Long Ashton Research Station, University of Bristol.

Medical Research Council SRS 297 *The Composition of Foods* by R. A. McCance and E. M. Widdowson. HMSO.

US Department of Agriculture, Agricultural Research Service, *Nutritive Value of American Foods in Common Units*, Agriculture Handbook No 456 by Catherine F. Adams.

US Department of Agriculture, Agricultural Research Service, *Composition of Foods*, Agriculture Handbook No 8, by Bernice K. Watt and Annabel L. Merrill.

2 How It All Started

More than 2,500 years ago, the people of China and Egypt buried their dead with the comforts necessary to help the deceased make their journey into the next world. While only effigies of pets and slaves were interred with the body, in the case of food they used the real thing, usually in the form of cereals and seeds.

Some of these seeds have since been sprouted by modern scientists. They had retained the precious germ of life because food spoilers—moisture, air, insects or predators and light—were automatically excluded by the tightly sealed tombs.

These ancient people had learned through observation that to preserve food in a safe and palatable form the first consideration must be the elimination of moisture. As long as water is removed, and is kept out, the food will keep. None of our modern poisons and chemicals, which we are assured are deadly to objectionable bacteria but somehow harmless to human body cells, were necessary.

Unfortunately only a very small, and probably unrepresentative, portion of the history of early peoples has come down to us. We do know that man was drying food long before he could write about it, but we have very little information about how he did it, what foods he dried and how they were stored.

In the period some 30,000 years BC we know that large areas of the northern hemisphere were scourged by hot, dry winds. Perhaps it was during this era that man first discovered that meat, if left outside in such conditions, remained fit to eat far longer than fresh meat. Certainly early man had to rely on nature to do his drying for him.

He cherished any seed grains, dried berries, nuts or any other naturally preserved foods he could garner and hoard for emergencies. So, when he became a farmer, it was a natural step to attempt to aid nature in the drying process. Most early drying must have been done out of doors with the sun as the only heat source. Caves, smoky, poorly ventilated and probably crowded, would not have encouraged much drying over the family fires.

All in all, early drying methods were probably time-consuming, inefficient and subject to constant attack from predators, including men from neighbouring tribes. Because the air flow and heat could not be controlled, these early sun-drying attempts were only relatively successful.

Wandering tribes must have found dried food a welcome addition to their diet, and perhaps even a necessity. As the climatic conditions changed, our ancestors were forced to migrate, and stronger tribes undoubtedly drove the weaker ones off the more productive and temperate lands. So the ability to carry light, nourishing food could have meant the difference between life and death while travelling over bleak mountain passes or through the parched desert.

Dates have for centuries been a staple food of the Middle East. In ancient times they were dried, then pressed and minced with a form of barley-paste to make a good tasting and long lasting food. Another custom was to extract the juice from dates, let it dry until it was the consistency of thick syrup, then use this whenever sweetening was required.

But it was not only in the warmer climates that food was dried and stored for the winter. In northern Europe, with its intense cold, heavy snowfall and shorter harvest periods, medieval communities often found that they had to kill their livestock because fodder for the winter was in such short supply. Salting or preserving meat in brine became popular, along with drying, to ensure at least some meat supply for the long dark winters.

Sometimes the meat or fish was placed in a salt pit, and it took a lot of hard physical labour to pound the hard chunks of coarse salt into a form fine enough to turn out a good salted food. Soaking in salt water was a fairly effective means of preservation, and when the Far East was at last open to European merchants, the addition of spices made the food much more palatable.

In Poland certain traditional dishes, such as cabbage with apples, and sauerkraut with caraway seeds and peppercorns, are preserved by an exotic blend of spices and salt, without the use of vinegar.

In England some food shops still display huge slabs of white salted and dried cod for sale in their windows, most of it imported from Spain and Portugal. Fish is still dried in much the same way along the northeast coasts of both Great Britain and the US.

Eskimos have survived for centuries on a diet of seal meat and fish. During the summer when fish could be caught in abundance, they would throw their catch on the beach to dry in the sand. Later, when food was scarce and all the fishing waters were frozen solid, this dried fish became a staple part of their diet.

Unfortunately the sand left on the fish often wore the Eskimos' teeth down to the gums, but in spite of this disadvantage there can be little doubt that this primitive method of drying fish provided much-needed food during the long dark months of the Arctic winter.

Not too long ago, in Mexico we saw the fishermen along the shores of Lake Chapala spread their catch on the roads to dry. They said the fish dry faster this way because the black surface of the tarmac reflects the sun's rays. How this may affect the flavour of the fish is another matter!

Primitive peoples seem to have always dried meat. Both the American Indians and some tribes in Africa use the following simple, but effective, method.

The meat, always lean if possible, is cut into very thin strips, then hung on a drying apparatus made of narrow pieces of rawhide stretched between poles, like a clothes-line. A fire is built underneath, and between the wind and the heat from the fire the meat is dried until it is bone hard, producing 'jerky' or, as it is called in Africa, 'biltong'.

This is just a little different from what we normally would call food-drying, for while the heat from the fire dries the meat, certain fat-soluble chemicals in the smoke are absorbed into the food as the moisture is carried away. The keeping quality, however, still depends more on the loss of water than on the chemicals from the smoke, so this could be considered as a flavoured, dried food.

Some tribes from the southwestern parts of the United States, living in harsh semi-desert terrain, even dried the beans from the mesquite tree, and when times were very bad, dried acorns as well. Both beans and acorns were sometimes ground or pounded into a flour although the flavour must have been rather bitter. Clint Seibert, of Vernonia, Oregon, a modern version of the mountain man of pioneer times, tells how the Indians used to gather wild Oregon grapes, crush them and form them into cakes. These cakes were spread on hot rocks to dry in the sun, then stored for winter eating.

Euell Gibbons met a tribe of Navajo Indians in Arizona who use the dried Rocky Mountain bee-plant for emergency rations. They first boil the plant down until most of the moisture has evaporated, then dry the residue in sheets for later use. They also find it serves as an admirable black paint, dissolving the sheets in a little water and using the resultant colouring to decorate their pottery.

Even animals seem to use food drying instinctively to help ensure a winter food supply. We all know how squirrels put away nuts for the winter, but, in a remarkable book by Sally Carrigher, *Wild Heritage*, we learn that pikas, a very small species of rabbit, gather stalks of plants like lupins, goldenrod and fireweed and spread them on flat rocks to dry. The seedheads are arranged on the outer edge of the heaps of seeds the better to catch the rays of the sun. When thoroughly dry, these tiny little animals gather up their 'hay' and store it carefully under rocks to help them survive during the winter.

But what possible interest, other than perhaps historical, does this ancient process of food preservation hold for modern man? It is easy to forget that it is less than a hundred years since Pasteur made his discoveries of the reasons for food spoilage. Up to this time man preserved food either by drying, or by storage under the ground or in cellars. Pasteur made possible the development of safe techniques for preserving food in bottles and tins.

Then in the middle of this century George Birdseye made frozen foods commonplace in millions of homes. The monotony of man's diet through the centuries, dependent on the seasons and his immediate environment, was relieved by exotic delicacies from the far corners of the earth.

But now, with oil prices rising and world food shortages threatening, the industrialized nations face a situation which the underdeveloped countries have lived with for years. These shortages, combined with spiralling food prices, threaten the health and peace of mind of everyone.

Dr E. F. Schumacher, author of the best-seller, *Small Is Beautiful*, in an article published in the *Soil Association Quarterly Review* (Sept 1975) says:

> The Western way of life, even in a much more modest form than its American model, requires the annual use of several tons of 'fossil fuel' per person. But the world's resources of fossil fuels, obtainable at a 'reasonable' cost, are strictly limited.

In the same issue a report from the 1975 World Food Conference in Rome said:

> US agriculture requires an unconscionable amount of energy. The energy spent in getting food to your mouth is about ten times the energy you get from it.

The report then goes on to estimate that, if the present rate of US energy consumption for agriculture and food processing and packaging continues, it would use up the equivalent of the entire output of the Alaskan Prudhoe Bay oil discovery in thirteen years.

British experts like Dr Kenneth Blaxter FRS, director of the Rowett Research Institute, Aberdeen, Scotland, and consultant director of the Commonwealth Bureau of Nutrition, believes the UK could become self-sufficient if people would change their diet, with a resulting change in agricultural policy. This opinion is shared by Dr K. Mellanby in his book *Can Britain Feed Itself?* (Merlin Press, 1975) who answers his own question in the affirmative. Dr Mellanby states that UK cereal production, for example is already adequate to feed the country without imports. The implication must be that, with proper planning and forethought, vegetables and fruits, in varieties that adapt well to the climate, could be grown in quantities large enough to help end much of Britain's food importation, with its disastrous effect on the balance of payments.

So home food drying is not simply an interesting process historically. With worldwide food and power shortages upon us, it makes good sense to find alternative ways to feed ourselves.

With the newly designed equipment now available to dry food at home, the process is far more efficient and enjoyable than ever before. Plastic bags and plastic containers also make storing the food much safer and easier.

I fully recognize the value of freezers as exciting and useful additions to the home-maker's kitchen. But they do consume great amounts of energy. Home bottling (or canning as it is called in the US) is for many people too time-consuming, and can involve too much heavy lifting for elderly or disabled people. Home food drying offers a practical, inexpensive and easy alternative, or complement, to the above methods of home food preservation.

More and more people are searching for viable alternatives to the present dilemma of high food prices, scarcity and expensive food preservation methods. After reading this book, work out a pilot plan for drying your food, and see for yourself what possibilities this ancient, but modernized, method holds for you.

3 How Food Drying Works

Before I actually started drying food, I found it helped very much to learn something about the underlying principles. Unfortunately there was almost no literature on the subject, and many of the so-called experts admitted there was plenty of room for disagreement about methods, and even more room for experimentation.

Nevertheless, there are certain basic facts that are vital to successful drying. If you absorb the simple principles in this chapter, you should then understand what you are doing with your dryer, and will end up with a much better product than if you start off half-informed.

You should first examine the reasons why food spoils. Moisture, essential to all life, allows the various types of organisms, which cause food decay, to flourish. So the first step is to remove as much moisture as possible.

Doing this requires two factors: heat and movement of air. Heat causes the evaporation of water from food, and the more water that is removed the better the state of preservation. (Most foods are between 75 per cent and 90 per cent water, which explains why your dried food shrinks sometimes to only one-fifth its original size.)

But merely driving out the moisture with high temperature is not the answer. If the temperature is too low, the food may spoil before it dries. If there is too much heat, the moisture will be driven out too rapidly and the outside surface of the food pieces will harden (this is called 'case hardening') and act as a barrier to further water loss. Then you might end up with food as dry as a bone on the surface but soggy, and often mouldy, inside.

Another problem with very juicy, sweet foods is that if the temperature is too high, especially at the beginning of the drying cycle, the food will drip and this means a loss of flavour, which should definitely be avoided.

So, what is too high or too low? Here we find differences of opinion among the few authorities who express an opinion on home drying of foods. When I first began drying food I experimented for myself, asked questions of anyone I could find who had ever dried food, read the few books and articles available on the subject and finally came up with a figure of between 32°C (90°F) and 43°C (110°F) for most foods.

Heat causes protein changes in food as well as affecting the aromatic oils which give flavour. If the temperature is too high you might even end up by cooking the food, which would give you a very tough and unappetizing product. It is quite true that commercial dehydrating methods do use higher temperatures, and that the food is dried in less than twenty-four hours. But the food industries have to turn out as much as possible to justify their huge investment in equipment and buildings. As the limited duration of most growing seasons forces these commercial plants to turn out enormous quantities of food every operating day, the higher temperatures are necessary.

But this is not a problem with home drying. Sufficient food can be dried without using these high temperatures. The single ideal heat range depends on the food you are drying and whether or not your time is limited. In the initial stages, the temperature inside the dryer may actually register much higher than our recommended 32°C–43°C (90°F–110°F), because the water in the food keeps the temperature lower by evaporative action.

Because of this, some commercial firms use two-step dryers, but this is quite unnecessary for home use. One reason there is often conflicting opinion among researchers on this subject is because for many years commercial methods were the only ones available for investigation. Unfortunately those methods suitable for large scale operations are usually not applicable nor indeed advisable for the home drying enthusiast.

As an example of how primitive some methods can be and still work, I quote from an instructor in survival techniques and home food drying.

> Any place you can feel warm air rising, fix a scaffold over it. Prop [it up] on boxes or chairs, and then stretch some cloth,

like cheesecloth or butter muslin, over it. Anything that will let the air through is fine.

Use a thermometer and try to raise the temperature to 38°C (100°F). Put the food on the cloth and scatter the pieces thinly so air can surround it. Never let your drying food touch metal. And remember, vitamins are destroyed if the temperature goes over 57°C (135°F).

This advice was given to a class of young Girl Scouts who were learning to dry their own food to take with them on outdoor survival exercises. The instructor himself now uses a box dryer, because he found it so much more efficient, but we quote him to show that even with such rudimentary equipment you can dry food satisfactorily.

As food dries it becomes more concentrated. Then the drying process slows down and the food is not kept as cool by evaporation. The final process is even slower. Water must, however, continue to diffuse from the centre of each piece of food. This is why, when the initial temperature is too high, or the initial drying air too dry, a hard impervious layer forms which interferes with the final state of evaporation. The centre of the food particle doesn't dry and this leads to mould, souring or rot.

Air Movement

Without the movement of air, the space around the food would become saturated with water vapour, no more evaporation could take place, and eventually the food would spoil. So in order to allow this free passage of air around each piece on the tray, they must be kept separate. This does not mean, however, that the drier the air, the better the results. In this too, there is a happy medium. The air is *recirculated* in the dryer, either by convection heating or by a small fan, and approximately 10 per cent new air is drawn into the drier. This cuts down on heating costs as well as improving the final quality of the dried food.

If a decision must be made between not enough air and too much, decide in favour of too much. We tried a dryer with a fan that was too small, and even with more heat it still took longer to dry the food.

Of course some common sense must be used. In drying herbs, for example, if the force of the breeze blows the leaves about, they should be covered with some suitable screening material like cheesecloth or old, clean net curtains.

For the above reasons we advise that food should never be packed or stacked on the trays. Stacked pieces of food will not receive the necessary circulation of air and this will delay the drying process.

Also the inside temperature of each dehydrator often varies and the air movement differs between trays inside the same dryer. So, with thick foods especially, you should consider shifting or rotating the position of the trays occasionally to be sure the food dries evenly.

Watch also, when drying halves of fruits like peaches, plums, pears, or apricots that you always place the *skin side down* on the tray surface. This not only keeps the fruit from sticking but also speeds drying time by exposing more of the cut surface of the fruit to the air flow. As the drying continues the food will shrink, so it is permissible to have the pieces touching, but they definitely should not overlap. About 1–1¼kg (2–2½lb) of fruit per square foot (77 sq cm) of tray space is the rule used in commercial drying, and it is a good one for the home dryer to follow.

Storage

Proper storage is essential for successful drying. It depends on eliminating the four enemies of dried food—air, moisture, light, and insect or rodent infestation.

Air Store dried food in small packets. Meal size servings are generally a good amount. Fill brown paper bags as full as possible, squeezing out the air, then seal tightly, using tape, rubber bands, string, or wire twists. Or you can twist the open end of the paper bag tightly shut, and dip in melted wax to seal. Then place in plastic bags, which should help to keep out weevils.

Store in large metal or plastic containers with tight-fitting lids. Possible sources of storage containers are hotels, restaurants, school cafeterias, sweet shops and bakeries. Ask for bulk containers for such foods as ice cream, sweets, flour, mayonnaise, and shortening. We found these valuable containers are usually discarded and can be had for little or no money.

In Great Britain the 'Victory V' cans with their press-down metal lids make excellent storage. They are metal and keep out the light. Sweet jars, unless stored in a dark place, are not quite as good, as light is definitely detrimental to the quality of dried food. However, if you first store your food in brown paper bags, this should keep out the light. Look for these points when selecting storage containers: tight fitting lids; they should be insect and rodent-proof, lightweight and light-proof if possible.

Dry Ice In order to live insects must have air. One way to make sure they do not infest your dried food is by the use of CO_2, also known as dry ice. Because it is a heavy gas, CO_2 will settle to the bottom of the storage container and force out the lighter air. It is especially useful if you are storing large quantities of food such as grain.

You can buy CO_2 either in cake form (used to keep ice cream cold for short periods) or in gas form in metal tanks. The latter is best for large quantities, but check with the shop where you buy the gas for instructions on how to use it.

For storing in 5-gallon (22l) cans we suggest the following procedure. Spread the dried food about an inch deep on bottom of can. Wrap 32 cu cm (2 cu in) of dry ice in a folded towel or newspaper, then crush into very small pieces, using a hammer or other heavy tool. (For smaller containers, use a little less dry ice.)

Spread ice over the first layer of food, then fill the container with the rest of the dried food, and put the lid on but not tightly. Leave the can to sit for about 30 minutes. Gas produced by the evaporation of the dry ice will escape. Then press the cover down, and check carefully for any bulges in the side of the can. If you see any, remove the lid but with great care. Wait for from 3 to 5 minutes then seal the lid, tightly this time.

Food stored in this way must always be placed in an upright position. CO_2 is, as we have said, heavier than air, and it can leak out if the cans are placed on their side or upside down. Because weevils have been known to enter food by way of the frost on dry ice, it is advisable to blow off any frost before crushing the ice.

Moisture Dried food tends to absorb moisture from the surrounding air. This can then cause spoilage to occur, even though the food may have been correctly dried originally. Therefore great consideration must be given to placing a moisture barrier around the food when preparing it for storage. Plastic bags, plastic film wrap, cans, glass jars or freezing containers, as well as ordinary brown paper bags can be used. If you use paper bags, as I recommend, it is advisable to then place them in plastic bags, before storing.

To draw out any possible remaining moisture, you may want to try one of the following desiccants: calcium chloride pellets (not readily obtainable in the UK); silica gel, available at most chemists or pharmacists, and supplied in 500g lots; diatomaceous earth, also available from chemists.

Place a small amount, wrapped tightly in a cloth bag, in the final storage container after the dried food has been individually packaged and sealed. Do not put desiccants in with the sealed food. It must not come in contact with the food itself. Any moisture left in the dried food will combine with the material in the cloth bag. The desiccant can then be removed, dried in the oven and reused.

If you have doubts about the moisture content of your food these measures are worth the trouble. To test for suspected residual moisture, leave the bag of desiccant in the storage container for a few days, then examine to see how much moisture has been absorbed.

Another excellent way to test for moisture is to store food in glass jars, protected from light, of course, and in a few days check to see if any moisture has condensed inside. This is easily visible and will tell you if you have dried the food sufficiently. If moisture should appear, spread the food on the trays and return to the dryer. Usually an hour or two is enough.

Light After the food is packaged, it is ready for final storage. Light has a deteriorating effect on dried food, and for this reason you can wrap each container in brown paper if it is not light proof in itself. If your final storage container is plastic you may want to wrap the individual packages in foil, as sometimes plastic will pick up outside odours. One very important precaution: never store food in large plastic containers, such as rubbish bins (new, of course!) in a garage. They may pick up the smell of oil or any chemicals you may keep there.

Because of its imperviousness to odours and also because of the protection it offers against rodents, metal is the preferred material for final storage. Rodents have been known to gnaw their way through heavy plastic. I stored food for over four years in metal rubbish containers and it kept perfectly. But whatever you choose, be sure it has tight-fitting lids.

To summarize: your storage for dried food should be proof against air, moisture, light, and rodent and insect infestation. Properly stored, the food should keep well enough to feed your great-grandchildren, if you don't eat it first. At least it will always be ready in case of an emergency.

4 Building a Box Dryer

In this chapter you will find complete instructions for building a slightly modified version of a dryer designed by the Oregon State University Extension Service. A complete materials list, with detailed plans, make it possible for even an inexperienced handyman to construct a satisfactory home dryer.

You definitely do not need a home workshop, complete with all the power tools available, for this dryer. What you will need is a saw, a hammer, a screwdriver, a tape measure and some help from a right angle. A chalk line is handy for marking the plywood, but not necessary. A heavy lead pencil and a straight board can make a good enough line.

If you have all the material together, marked up and are ready to start work, you might consider renting a power saw. You will save yourself a lot of tedious hand sawing if you do.

The next chapter will discuss other types of dryers and how to build and use them. However, we believe this dryer to be one of the best models for size, results and comparatively low cost.

Materials List

2 sheets 6mm (¼″) plywood 122×244cm (4′×8′)
2 sheets 12mm (½″) insulation board 122×244cm (4′×8′)
1 piece 12mm (½″) asbestos board 46×76cm (18″×30″)
or
1 sheet of tin, same size
7m (24′) timber 5×5cm (2″×2″), in lengths divisible by 122cm
14m (48′) timber 2·5×2·5cm (1″×1″)
27m (90′) timber 2·5×2·5cm (1″×1″), in lengths divisible by 6
Screening for trays: 7 pieces 76×46cm (30″×18″)
Note: In Great Britain, Netlon plastic non-toxic mesh screening. In US, fibreglass screening is recommended.
10 lamp receptacles or porcelain sockets

9 light bulbs, 100 watts each
(or heating element, cone style if available)
6m (20′) #14 or larger asbestos insulated copper wire
1 cord connector cup
1 dairy thermometer
1 thermostat (optional, but definitely an advantage)
1kg (2lb) nails, 5cm (2″) size (US sixpenny)
1 electric fan, 25cm (10″) diameter or larger
contact cement
staple gun (optional but recommended)
power saw (same)

Timber should be marked and cut according to the following parts list. You will notice that all measurements take into account the dimensions of finished timber, so a 5 × 5cm (2″ × 2″) is actually 4cm (1⅝″) square. If your timber is not finished to these dimensions, then slight changes in the notching will have to be made.

Also, if you decide to build a smaller dehydrator, then change the measurements on this list as you wish. Since all parts are numbered this should not present any problems.

Parts List

Part No	Amount required	Name of Part	Type of Material and Size or Amount
1	2	Sides	6mm (¼″) plywood 402 × 107cm (40″ × 42″)
2	1	Back	6mm (¼″) plywood 55 × 105cm (21¾″ × 41½″)
3	1	Top	6mm (¼″) plywood 55 × 109cm (21¾″ × 42¾″)
4	1	Bottom	6mm (¼″) plywood 46 × 119cm (18″ × 42½″)
5	1	Top door	6mm (¼″) plywood 58 × 57cm (22¾″ × 22½″)
6	1	Bottom door	6mm (¼″) plywood 43 × 55cm (17″ × 21¾″)
7	1	Front stiffener	6mm (¼″) plywood 20 × 55cm (8″ × 21¾″)
8	2	Sliding air controls	6mm (¼″) plywood 15 × 15cm (6″ × 6″)
9	1	Lamp receptacle base	6mm (¼″) plywood 30 × 45cm (12″ × 17¾″)

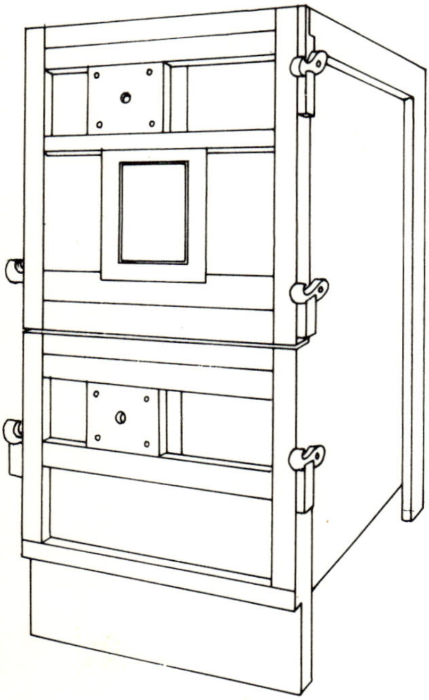

Fig 1 Dryer cabinet exterior view

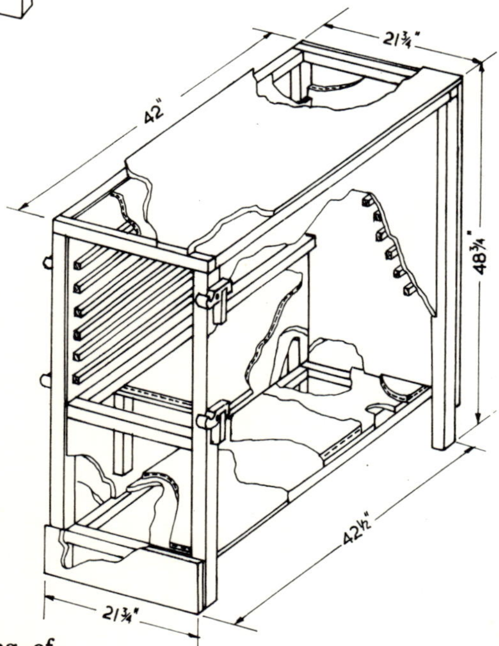

Fig 2 Cut away drawing of dryer to show its structure

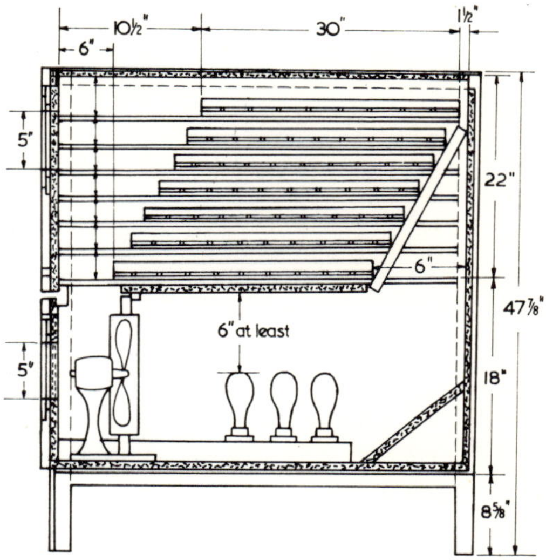

Fig 3 (above) Vertical section through centre of dryer Fig 4 (right) Top and bottom doors of dryer showing the variable air vent

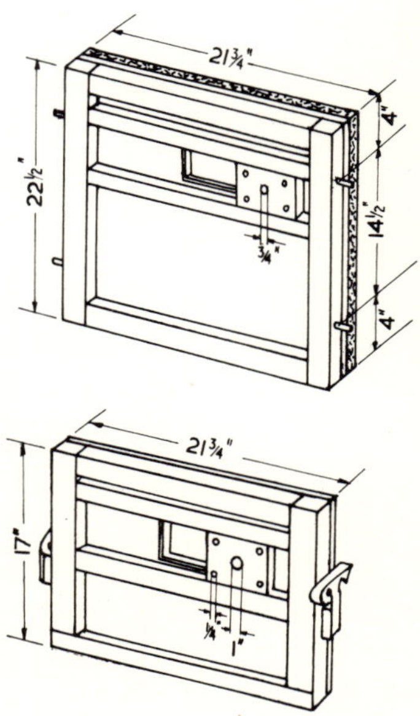

Part No	Amount required	Name of Part	Type of Material and Size or Amount
10	1	Fan base	6mm (¼″) plywood 20 × 45cm (8″ × 17¾″)
11	2	Sides	12mm (½″) insulation board 1 × 1m (39¼″ × 39¼″)
12	1	Back	12mm (½″) insulation board 47 × 97cm (18½″ × 38⅛″)
13	1	Top	12mm (½″) insulation board 46 × 104cm (18″ × 41″)
14	1	Bottom	12mm (½″) insulation board 46 × 105cm (18″ × 41½″)
15	1	Top door	12mm (½″) insulation board 55 × 57cm (21¾″ × 22½″)
16	1	Bottom door	12mm (½″) insulation board 41 × 45cm (16″ × 17⅞″)
17	1	Bottom tray baffle	6mm (¼″) asbestos board 46 × 76cm (18″ × 30″)
18	1	Corner baffle	12mm (½″) insulation board 30 × 46cm (12″ × 18″)
19	1	Fan diaphragm	12mm (½″) insulation board 38 × 44cm (15¼″ × 17½″)
20	4	Legs	5 × 5cm (2″ × 2″) dressed 117cm (47″) long
21	2	Top side rails	5 × 5cm (2″ × 2″) dressed 102cm (41″)
22	2	Top end rails	2.5 × 5cm (1″ × 2″) dressed 55cm (21¾″) long
23	1	Door stop	2.5 × 5cm (1″ × 2″) dressed 46cm (18″) long
24	2	Bottom end rails	2.5 × 5cm (1″ × 2″) dressed 47cm (18½″)
25	2	Bottom side rails	2.5 × 5cm (1″ × 2″) dressed 102cm (41″)
26	2	Top door framing	2.5 × 5cm (1″ × 2″) dressed 57cm (22½″)
27	2	Top door framing	2.5 × 5cm (1″ × 2″) dressed 47cm (18½″)
28	2	Bottom door framing	2.5 × 5cm (1″ × 2″) dressed 42cm (17″)
29	2	Bottom door framing	2.5 × 5cm (1″ × 2″) dressed 47cm (18½″)
30	2	Heating unit runners	2.5 × 5cm (1″ × 2″) dressed 76cm (30″)
31	4	Top door latches	2.5 × 5cm (1″ × 2″) dressed 7.5cm (6″)
32	2	Hook and eye latches	
33	14	Tray supports	2.5 × 2.5cm (1″ × 1″) dressed 102cm (41″) long

Rotating trays in the smaller dryer

Onions half-way through the drying process

Onions completely dried

Part No	Amount required	Name of Part	Type of Material and Size or Amount
34	2	Fan diaphragm stops	2·5 × 2·5cm (1″ × 1″) dressed 32cm (13″) long
35	1	Fan diaphragm stops	2·5 × 2·5cm (1″ × 1″) dressed 46cm (18″) long
36	14	Tray sides	2·5 × 2·5cm (1″ × 1″) dressed 76cm (30″) long
37	2	Baffle slides	2·5 × 2.5cm (1″ × 1″) dressed 76cm (30″) long
38	14	Tray ends	2·5 × 2 5cm (1″ × 1″) dressed 46cm (18″) long
39	4	Butt hinges	7½cm (3″) each

If you are not too familiar with carpentry, or reading drawings, then start with the doors. This means reinforcing the plywood sheet at the edge with 2·5 × 5cm (1″ × 2″) timbers, then fitting the guides for the fresh air control. These guides are two pieces of lath. One acts as a track for the sliding air control, and the other overlaps to keep the air control in place.

For the beginner, the following condensed part list may be of help.

Name of Part	Part Number
Sides	1, 11
Back	2, 12
Top	3, 13, 21, 22
Bottom	4, 14, 24, 25
Top door	5, 15, 23, 26, 27, 28, 29, 31, 32
Bottom door	6, 16
Front stiffener	7
Air controls	8
Heating	
Lamp receptacle base	9
Runners	30
Fan	
Fan base	10
Fan diaphragm	19

Fan stops	34
Fan stops	35
Bottom tray baffle	17
Corner baffle	18
Baffle slides	37
Corner baffle braces	38
Trays	
Sides	36
Ends	40
Screening	41

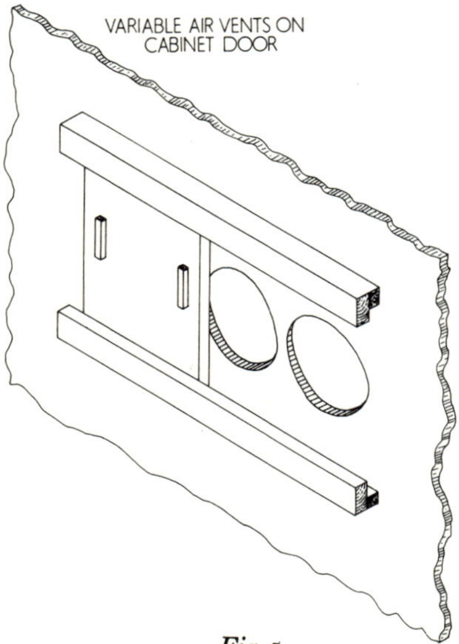

Fig 5

In some parts of the country special fans with slow speed, and heat cones that screw into an ordinary light socket, are available. These can be used to good advantage in a dryer, and the heat cone is often substituted for the light bulbs. Also, if you can find a bathroom heater of 1,200 watts and with a slow fan, this will do a good job and save you a bit of work.

If you have a router, make a 1cm ($\frac{3}{8}''$) slot in one edge of the guide.

Or, you can use a 2·5 × 5cm (1″ × 2″) timber as the guide, and a short piece of 1cm ($\frac{3}{8}''$) quarter round to hold the air control in.

The next step is to tack on the 1·2cm ($\frac{1}{2}''$) insulation, and you have finished the first main part of the dryer.

It should be noted here that construction could be simplified by using 2cm ($\frac{3}{4}''$) plywood, and thereby eliminating the 5 × 5cm (2″ × 2″) timbers, and the insulation board. However, this would be more expensive, and would result in a heavier unit. If you decided to use the 2cm ($\frac{3}{4}''$) plywood, it would only affect the material for the outer shell. Looking at the plans, you can see that this would mean eliminating the 5 × 5cm (2″ × 2″) timbers, the 6mm ($\frac{1}{4}''$) plywood, the insulation board, and some of the 2·5 × 5cm (1″ × 2″) timbers. I have seen dryers built this way, but personally I prefer the lighter unit.

Your next preparation should be the sides. (1) This board is 102cm (40″) high and 107cm (42″) long. The top edge should be notched 2cm ($\frac{3}{4}''$) by 4cm ($1\frac{5}{8}''$). After notching, measure and mark the position of the tray supports. (33) Measure from the top, starting with 10cm (4″) then 6 more measurements of $7\frac{1}{2}$cm (3″), remembering that one side faces the other. Then match the top of the tray supports to the marks and nail. See illustration No 3.

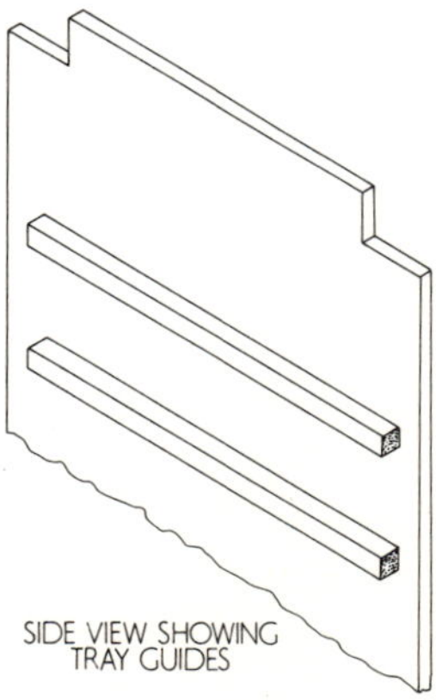

Fig 6

If you wish to nail from the plywood side through the 2·5 × 2·5 cm (1″ × 1″), either use a chalk line to mark the opposite side and nail along the line, or if you have no chalk line you can nail through the runner so the end of the nail comes out of the plywood. Do this at both ends.

Tie a string between the nails and nail along the string. Then if you wish, you can pull out the original nails.

The legs are nailed to the ends so that the top of the legs is even with the bottom of the notch.

Next, nail the top side rail (21) on the side panel. It should rest on half of the top of the legs, but not encroach upon the notched area at the corners.

The bottom side rail (25) is next. This is mounted on the legs on the same side as the plywood. It fits up snug against the bottom without overlapping. Here again, room should be left at each end to fit in the bottom end rail. This may appear to leave too much looseness in the bottom of the 1cm ($\frac{1}{2}''$) plywood, but the insulation both on the outside and the inside will adequately correct this. So do not worry about fastening the bottom of the plywood side.

Nail the tray stop (40) on either side, over the tray slides. This will stop the trays so the air will reach each tray evenly.

The baffle (17) can be either heavy tin or asbestos. Slides for this baffle (37) are mounted below the last tray slide. It should be 15cm (6″) from either end and spaced to the thickness of the baffle plate. Sliding the baffle plate between the guide and the bottom tray guide should be a tight fit.

You will notice that the bottom tray does not have as much air circulation as the upper six. Being closer to the heat source seems to balance out this apparent lack so that even drying will result. If the baffle slides are a good enough fit, then later a bead of glue is all that will be needed to hold the baffle in place.

You are now ready to form a box of your sides. Take the top end rails (22) and the bottom end rails (24) and fasten them to one side of the dryer, at right angles to the bottom. If your plywood end and top and bottom are cut square, then the squareness of these braces is not critical, as you will correct any error when you attach the plywood.

Attaching the second side might give you a little trouble. It is easier if you have someone to hold them upright. Then it is just a matter of fitting the ends in their right place and nailing them. If you are doing this alone, and want to go ahead with the project, then you can either temporarily tack one side upright to a bench or a wall, or you can lay the unattached side on the ground and lift the other side over it and then attach it.

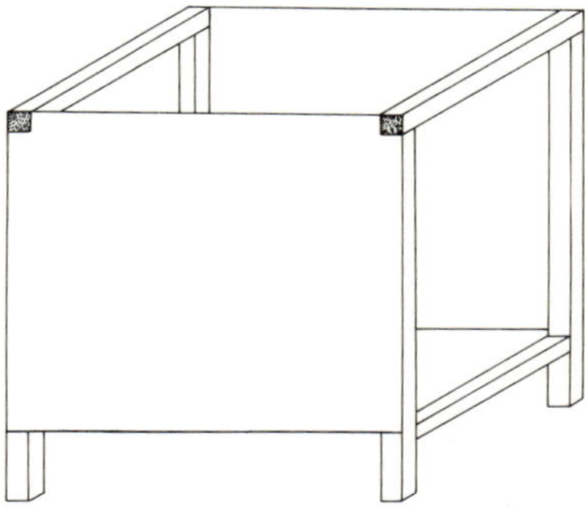

CABINET ASSEMBLY *Fig 7*

The plywood bottom (4) should come next. If this fits in and matches at all edges, you can be proud. If not, nail one side, then push on the other side until it is even. If you can get the bottom plywood to fit in well, so the bottom is square, you might want to paint the bottom rails with contact cement and then tack on the plywood. This will give added rigidity to your dryer.

The back plywood sheet (2) goes on next. Stand the box on the front, and it should be easy to apply. Here again, glue under the plywood will make it more rigid. Now add the front stiffener (7).

Leave the top until last, because the corner baffle (18) has to be put in.

Attach the back insulation board (12) and the bottom insulation board (14) in that order.

Measure 22½cm (9″) from the back along the inside of the bottom.

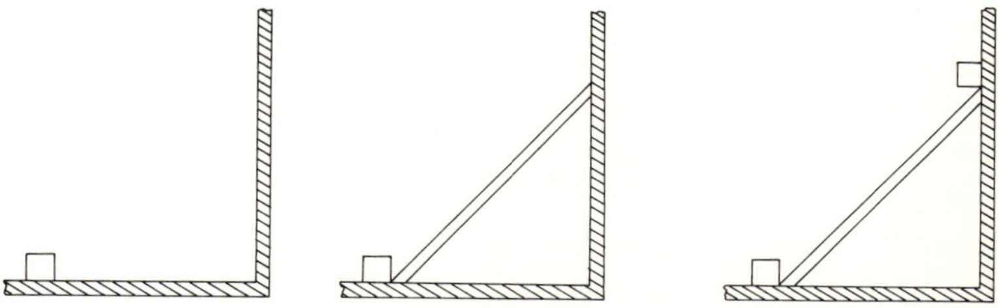

BACK CORNER BAFFLE CONSTRUCTION

Fig 8

At this point nail a corner baffle bracket. Fit the corner baffle against this bracket. Fasten the top bracket.

You may want to do this with screws from the back. In this way, no nails will be showing.

Now slide in the baffle and glue it.

Add the top (3) and then the top insulation (13).

Add the door stop (23). Drill holes if you are afraid of splitting the wood, using a bit about half the size of the nail.

Attach the doors. Sometimes it is easier to nail the door in place without driving the nails in all the way, and then attach the hinges and hook and eye. Then pull the nails. This should give the easiest and tightest fit.

The major work is now done, and you are ready for the electrical components. The plan calls for light bulbs, but you can use heating elements which screw into lamp sockets. In either case, light bulbs or heating elements are hooked up in parallel as shown in the schematic overhead picture.

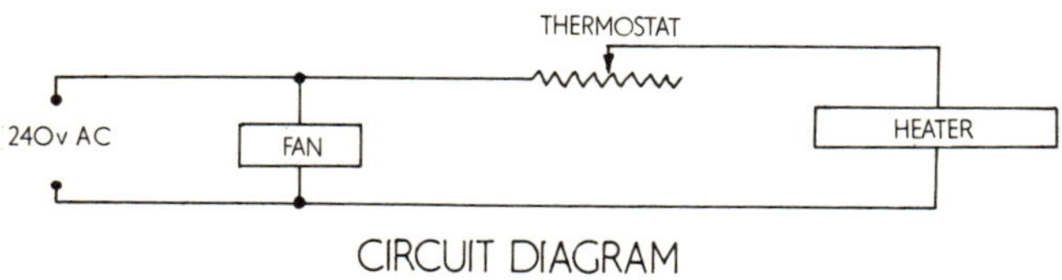

Fig 9 (NB Voltage used in USA is 110, not 240)

You will notice that the thermostat turns the heating off and on but leaves the fan running.

The fan diaphragm must be cut out around the fan and must fit against the bottom tray baffle. This ensures circulation of air through the drying chamber. The thermostat can be attached to this baffle or any place where it can be adjusted from the outside.

The thermometer can be hung on the inside, or you can drill a hole in the top and let it hang through to the inside.

The thermostat should be set by bringing the dryer up to the desired temperature as shown on the thermometer and then made to shut off the heating. This is much easier to do when you use light bulbs for heat, instead of heating elements.

Mount the components on a tray which slides in, seating against the baffle (17). Notch a place for the electric cord in the bottom so it can be taken out between the door and the bottom. This makes it so you can remove the complete unit, depending on where you mounted the thermostat.

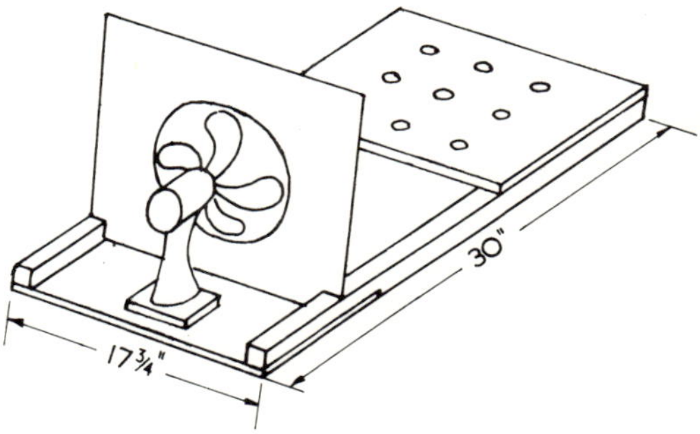

Fig 10 Fan and heating unit

Take the tray sides (36) and the tray ends (38) and construct your trays. You can use any artistry you want in making the corners, but being lazy, I just used butt ends. First I predrilled the holes, then dipped the butt in contact cement, and nailed them together. The screen was mounted on this (see Fig 10).

When using hardware cloth, some allowance should be made for the thickness of the wire so the tray will slide easily. I personally prefer a netlon or fibreglass screen, even though it will stretch a little, because it is easier to clean. You may want to put in a middle brace to support the screening.

A staple gun makes it much easier to fasten the screening to the frames, and it is even worth renting one just for this.

To make the trays sturdier, take some plywood scraps and cut out 28 triangles, measuring 3″ on either side of the right angle. Fasten these on the bottom after you have attached the screen to the frame.

If the whole tray bit seems like too much work, you can sometimes buy prefabricated window screens made to almost any size. These are generally aluminium and have a nice rigid aluminium frame. Check the price—they may not cost much more than you would pay for materials to make them yourself.

5 What Kind of Dryer?

In this chapter we will consider various methods of drying and describe the type of dryer used for each. Your individual choice will probably depend on such factors as how much space you have available, cost, whether the dryer will be indoors or outside, and if indoors, in what part of the house it will be used.

It should always be in a well-ventilated area so moisture evaporating from the drying food will have a chance to escape from the room. Having said this, I must admit that I kept one dryer against the wall in my front hall, dried enormous quantities of food with it, and never had a bit of trouble with condensed moisture. But as a general rule, try to have free circulation of air around the dryer, especially where the air vents are located, to avoid unnecessary problems with dampness.

You may want to start with just a small dryer, such as the one described on page 57, which is used over a heat register and has no fan. A solar dryer is practical if you live in a climate where dry, sunny weather can be counted on during the harvest season.

The best dryer we have found is the box dryer. It is a little more trouble to make, but it will do the most efficient job, and the model in Chapter 4 has a capacity of from 11 to 18kg (25–40lb) of food. The same idea can be modified to a smaller model if you find you don't have enough room for the larger size.

Plans for this smaller cabinet (see jacket illustration) are included because it has several advantages over the dryer in Chapter 4. Because it is smaller, it will fit into a house or flat with limited space. It looks like a piece of furniture and therefore is inconspicuous and can be placed almost anywhere in the home.

It uses a fan and a heater with 1,200 watts capacity, and has done a creditable job of drying for me over the years. I especially liked it because it serves as a secondary airing cupboard (this is a convenience not found very much in the US) to finish drying damp clothes during the winter months.

The trays I bought from a manufacturer. I also made regular framed screening trays, which worked quite well. These can be constructed in the same way as those for the box dryer.

This particular dryer is a good project for someone who enjoys producing something both beautiful and practical.

Parts List for Small Box Dryer

1 sheet Melamine, 2·4×1·2m (8′×4′)
12 joints
2 pair hinges
2 knobs
1 door catch
2 vents
1 1,200 watt heater with fan
thermostat
sheet of aluminium, 81×91cm (32″×36″)
Optional: framing and netlon screening for 4 trays
Approximate cost at time of publication:
Without purchased trays: £38 ($66·12)
With purchased trays: £50 ($87)

We made a dryer from a popular British clothes dryer which we found amazingly good (pages 51–2). It is comparatively inexpensive, lightweight, and multipurpose. You can dry your clothes in it as well as your food!

You will have to make your own trays, but this is a simple enough matter. We used netlon screening, of a very fine mesh so food particles would not fall through, tacked on to a wooden frame. The trays can be made in a very short time and are light, sturdy and washable. You should put a brace across the centre to keep the screening from sagging when drying heavy fruit like peaches or apples.

Because it has no fan it is not as efficient as the box dryer, but I dried potatoes, onions, apples and mushrooms with it and they all turned out perfectly. Be sure to use a thermometer with it though, as it has a tendency to rise a bit higher in temperature than it should. You can regulate the temperature by opening or closing the cloth covering.

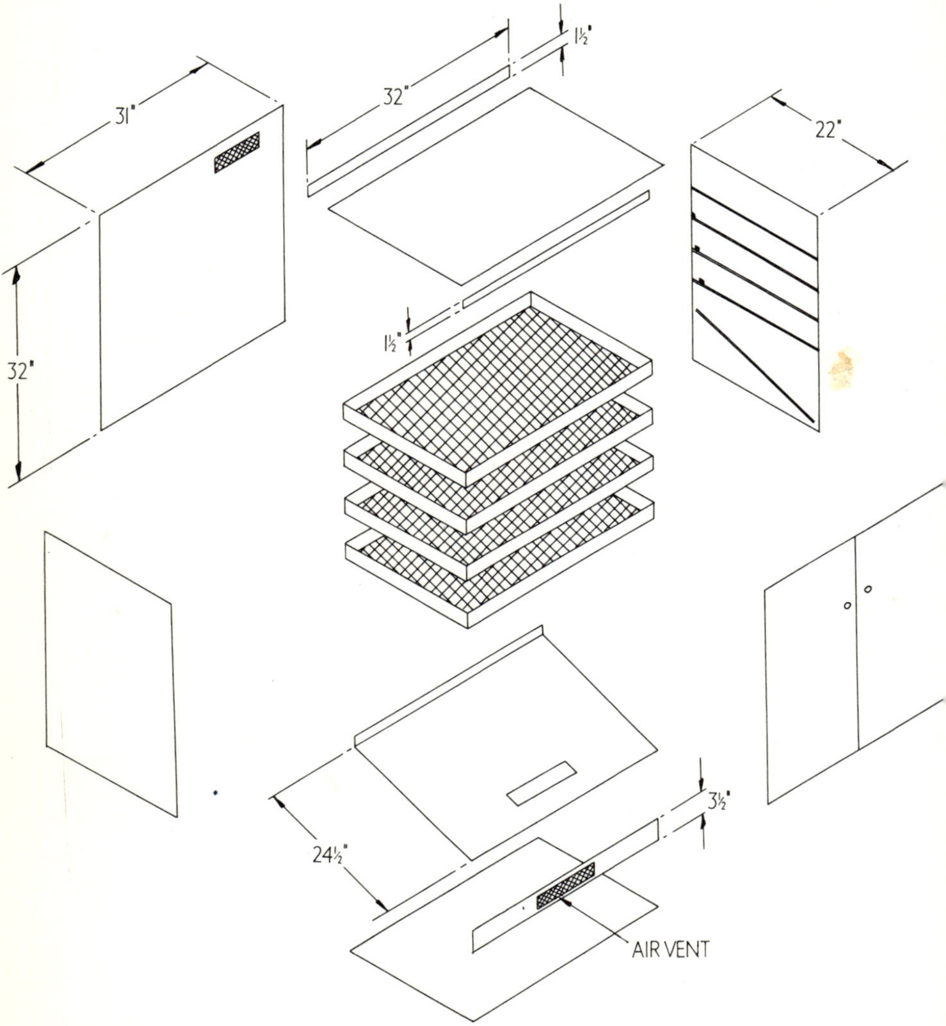

Fig 11 Expanded diagram of smaller cabinet dryer

Due to lack of air circulation, the temperature varies a bit more from one shelf to the next. Check this during the drying cycle, and rotate the shelves as needed. I did this and found it took very little time, and the results were ideal. The bottom tray, which is placed over the heater, seemed to get the hottest, but nothing scorched and changing the trays was a simple matter.

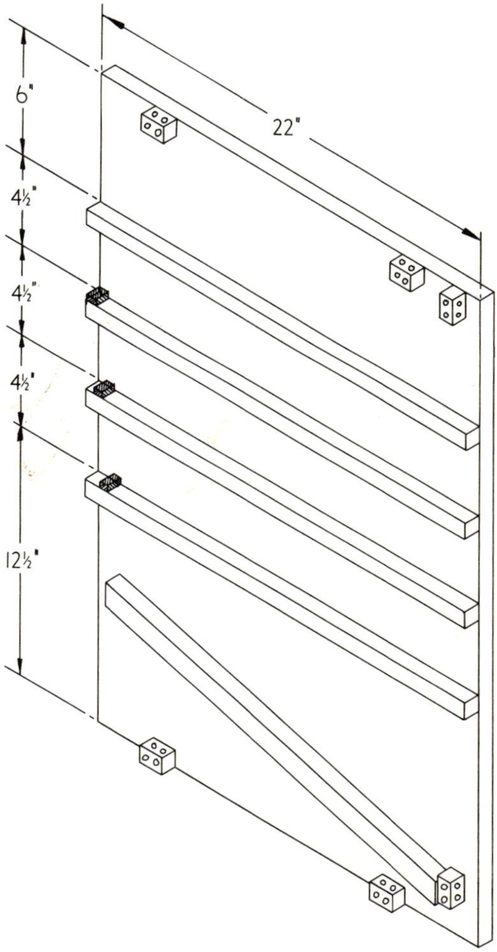

Fig 12 Side of cabinet, inside view, showing tray slides, stops and corner blocks

Sun Drying

While sun drying may sound like the natural and therefore the preferred method of drying, it is important to remember there is more vitamin loss in sun dried fruit. And the labour factor is much greater. Preparation of the fruit is just as exacting as for any other drying method. But since sun dried food will seldom dry in one

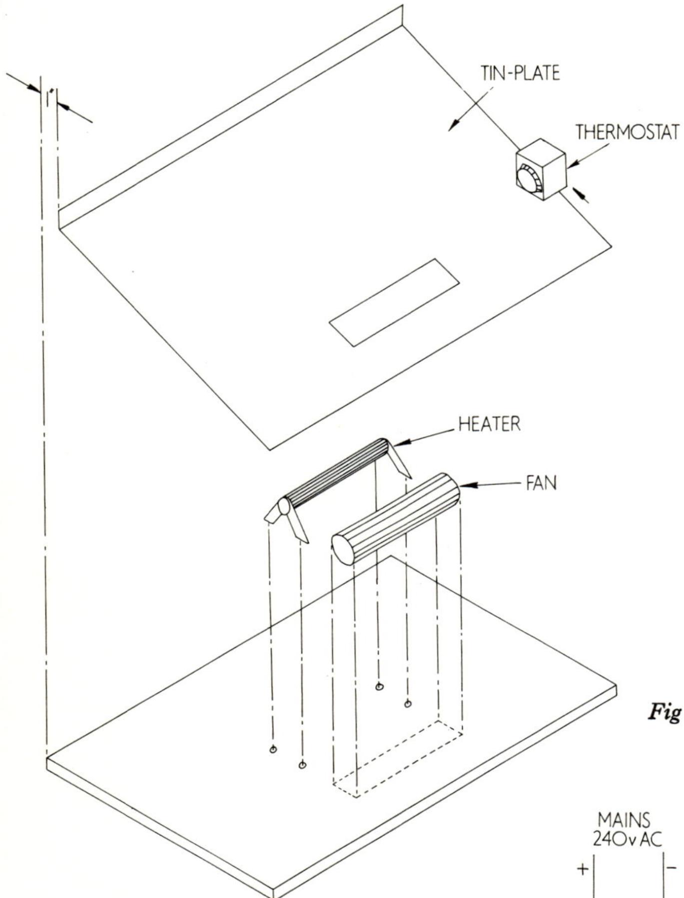

1"
TIN-PLATE
THERMOSTAT
HEATER
FAN

Fig 13

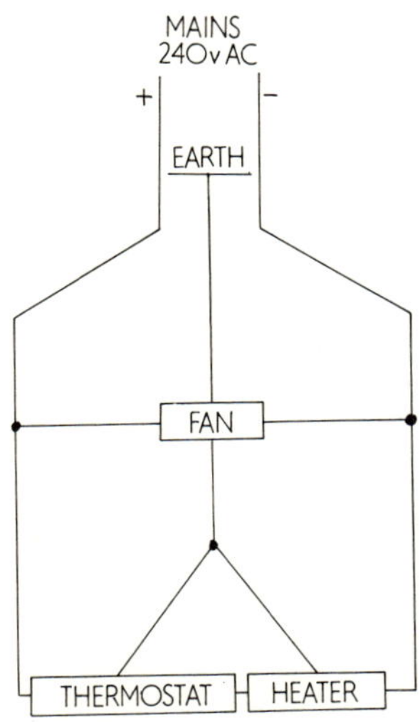

MAINS
240v AC
+
-
EARTH
FAN
THERMOSTAT
HEATER

Fig 14

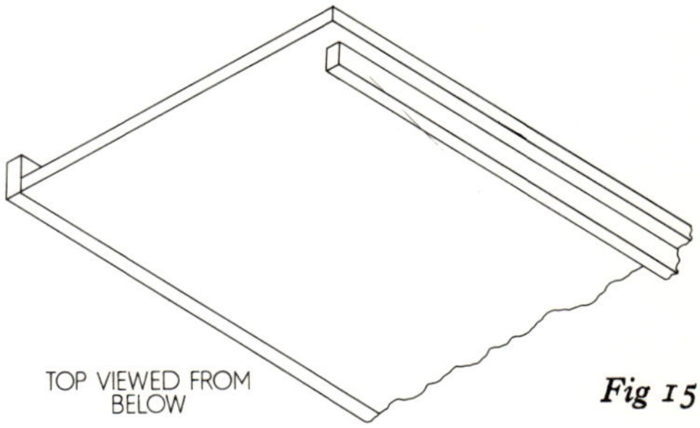

Fig 15

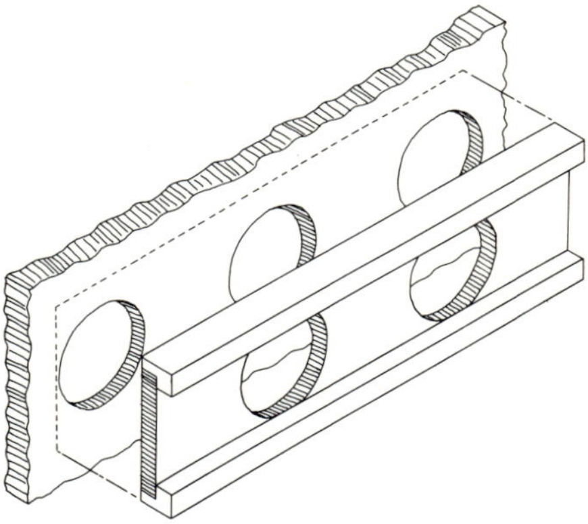

(*left*) Detail of air vents
(*below*) Mesh drying tray

Fig 16

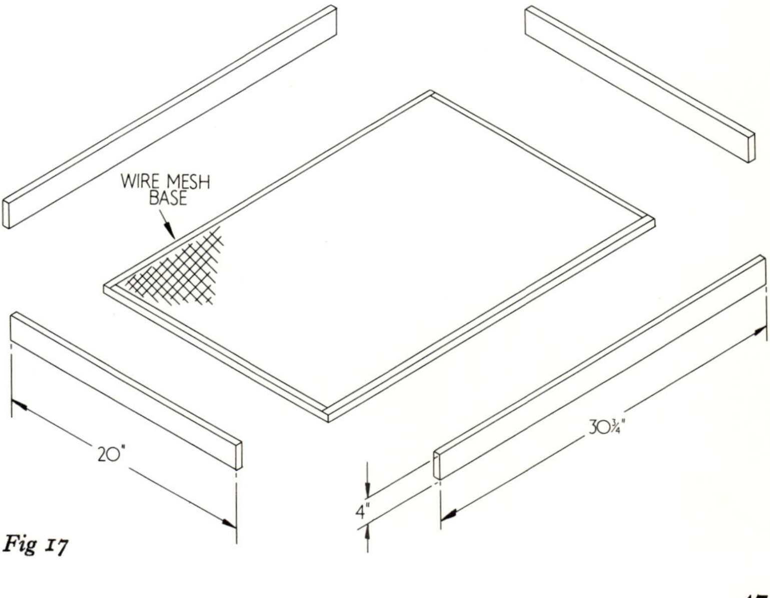

Fig 17

day, the trays must be moved indoors every night, or else covered for protection against moisture.

Sometimes an unexpected summer shower will spoil the fruit. Then there is the added problem of protecting it from birds, insects and dust. This is usually accomplished by draping the trays with a material like cheesecloth or old, clean sheets. Also, screened or glass trays can be put over the food, but this will cut down on the heat from the sun and lengthen the drying time.

A type of garden netting enclosure, which is very popular in the UK for protecting crops from birds, would be suitable. It might also make it a little easier to protect against showers as you could then fling a large sheet of plastic or tarpaulin over the whole dryer without worrying about injuring the food. This netting is made of either plastic mesh or chicken wire, so it is light and easy to handle.

When we tried to sun dry cherries (which incidentally end up like raisins, with a delicious sweetness and texture) the bees made it impossible to approach the fruit during the day. If the covering blew off, for example, we had to abandon that tray to the birds, rather than face a cloud of hungry, hard working bees.

It is all the extra labour involved in sun drying that helped make canning and freezing so popular. But if you live in a land of constant sunshine, low humidity during the growing season, and away from smog and traffic fumes, the following information should help you get good results.

As for any drying, use only top quality food. Prepare as directed in Chapter 6. Be sure the food is cut in small, even particles. Place on racks which allow a free flow of air. Even with sun drying it is the air which carries away the moisture in the food.

For this reason, unless you are drying very small amounts and are willing to take the trouble to stir the food and move it around frequently to keep it in the direct rays of the sun, we do not recommend using boards, doors, etc on which to spread the food. It will take much longer to dry unless there is air circulating around and through the drying surface.

If possible, set up your dryer near your house. You will have to make frequent trips to it during the day to stir the food and possibly to rotate the dryer. But keep it away from anywhere where cars or trucks might stir up unwelcome dust. The dryer must also be protected from animals, both wild and domestic, if you happen to live out in the country.

A roof makes an excellent location if it has safe, easy access. Also, heat reflected from the roof hastens the drying.

Suspending the racks from a clothes-line by wires has been done with some success. This provides good air circulation and relative safety from animals. You may have to experiment with the length of the wires and the angle, so that the food will get the full heat of the sun, but will not slide to the bottom of the tray as it dries.

When the weather is very hot it may be advisable to put the dryer in the shade for the first couple of hours. This will prevent the hard outside crust or case from forming. If you are unfortunate enough to have heavy mist or rain interrupt the drying, it is best to finish in an indoor dryer or in the oven.

Cold frames made with screening to allow air circulation can be used. See Chapter 4 for directions on how to construct the frames. Instead of using glass, as in a cold frame, you can construct a unit of two trays, one to be filled with food for drying, and the other inverted on top as a cover. This will keep out birds and limit, although not eliminate, dust and insects.

Fasten the trays together with hinges at the back, and use screen door hooks and eyes for the front. One suggestion is to use wood screws, instead of nails, for added strength. This is probably a good idea, as the solar dryer will be moved around much more than a box dryer.

The top screen must lift easily for convenient access to the food on the bottom screen. You might want to fasten several sets of screens together to make one larger unit if you plan to dry food in quantity. In this case we recommend putting the unit on rollers.

For any type of tray, regardless of which type of dryer you decide to make, use glue as well as nails for added strength. Basically a 2·5 × 2·5cm (1″ × 1″) frame, with butt corners, will be solid enough.

Sometimes the bottom of the tray presents problems. It is difficult to get fibreglass screening tight enough to support heavy fruit, such as peaches. Galvanized screening, 6mm ($\frac{1}{4}$″), seems to work best. *Never* allow ungalvanized material to come in contact with food as it causes a harmful chemical reaction. And if you sulphur fruit before drying (see Chapter 7), even galvanized screening should be avoided. The sulphur will react to produce undesirable chemicals.

An annoying problem sometimes occurs in drying sticky food—it sticks to the trays, which are then hard to clean. This can be prevented by applying a very thin coating of vegetable oil or beeswax to the surface of the tray before placing the fruit on it. Be sure to dry the food immediately after applying the oil, however, to prevent any possibility of rancidity developing. This will spoil the flavour.

During the drying process, stir the food occasionally to ensure even drying, and as mentioned previously, rotate the trays so they will catch the full rays of the sun. Do not leave trays out during periods of high humidity. Bring them indoors, or cover with plastic, a tarpaulin or a sheet of tin, as soon as the sun begins to set. Do not uncover until the sun has been shining at least two hours. When the food feels fairly dry, but has a bit of moisture left, move the trays into the shade and allow to dry a little longer.

All sun dried food should be pasteurized—that is spread on trays or baking sheets and placed in a 79°C (175°F) oven (the lowest possible gas setting) for 10–15 minutes. This will kill any possible insect infestation. Be sure the trays are protected from insects during the cooling periods. Store immediately. See Chapter 7 for details.

We found an interesting way to dry herbs, using discarded nylon pantyhose. We used to pick peppermint, which grows wild in many areas of the US and Britain, for a delicious and refreshing tea. After stripping the leaves from the stalk, we stuffed them into the legs of the tights and hung them in a warm, windy spot. They dry very quickly if the weather is right, but be sure to bring them in at night if it looks like rain, or if you usually have a heavy dew.

When crumbly to the touch, the herbs are dry enough. Strip the leaves from the stems which should then be discarded. Pack the leaves in any container which can be tightly covered, and store in a dark place, since light causes all dried foods, especially herbs, to lose flavour and colour. Coffee tins, thoroughly scrubbed to get rid of the coffee aroma, make ideal storage containers for herbs.

The clothes dryer uncovered showing onions laid out to be dried

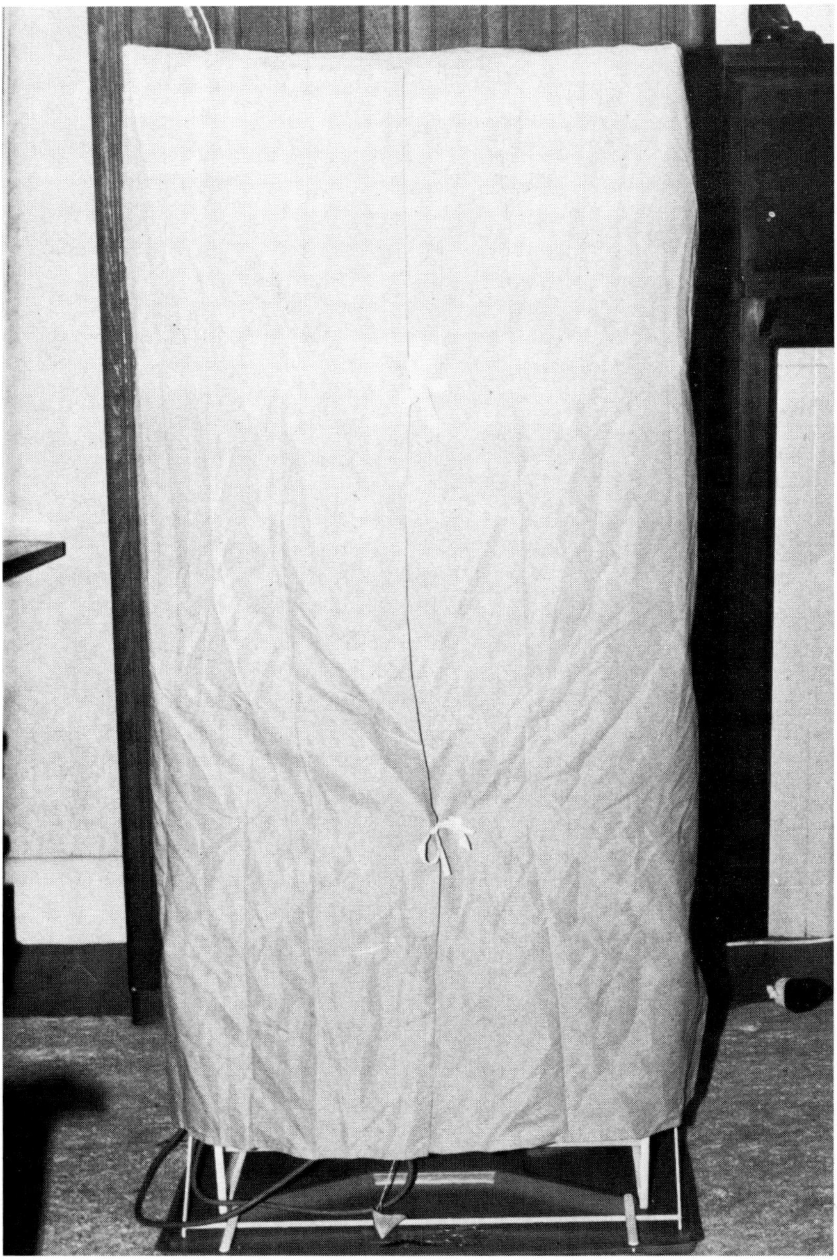

The clothes dryer in operation. Moisture-saturated air escapes from the open space at the bottom

This idea works well when you are in between drying sessions and don't want to start up the dryer for just a couple of trays' worth of herbs.

If you are lucky enough to live in a region with long periods of rainless, sunny days, you might try some of the following ideas submitted by a very successful lady organic farmer.

In winter when she dries olives, or if there is a very wet spring for the apricot harvest, she makes a tent out of a big sheet of clear plastic and spreads it over the drying trays. This has a sort of hothouse effect so if you do this, be sure to provide ventilation at both ends to prevent the food from mouldering. At the same time be sure birds can't get in or you will not have much food left when they finish!

She solves this problem by tacking the edges of the plastic to a large framed piece of screening. Any type will do, as long as the mesh is small enough to prevent birds from getting in.

If you use trays made of screening, rather than wooden slats as some farmers use, the fruit will dry faster if you place brown or white paper under the trays to reflect the sun's rays. Tinfoil or aluminium foil will have the same effect.

When the fruit is rubbery dry, it is placed in the oven at or below 65°C (150°F) to sterilize any insect worms or eggs. The food is then kept in the freezer until sold, usually within six months.

This particular grower adds a small amount of honey to her figs before drying which she thinks helps to keep them from going mouldy. It also adds to the flavour. She has tried this with apricots but says it depends on the individual's sweet tooth. There may be dietary reasons, too, for avoiding additional sweetening.

Converted Refrigerators

Old refrigerators can be converted into remarkably efficient and practical dehydrators, as can old freezers. As the drawings show, it takes very few modifications to convert a refrigerator on its way to the scrap heap into an inexpensive and quite satisfactory home dryer.

This kind of dryer is often kept outdoors, either in a garage, on a back porch or in a location close to the kitchen door for convenience. Because of the superior insulation, and because heat is so well reflected off the (usually) white interior, the only heat source needed is an ordinary light bulb, from 150 to 250 watts, depending on the outdoor temperature where the dryer is located.

You can use either the same shelves that are already installed, or fasten pegboard on each side and make sliding shelves, using 6mm ($\frac{1}{4}''$) bolts, along the same lines as those used in the box dryer in Chapter 4.

For this type of dryer you can keep the temperature up to 60°C–65°C (140°–150°F) without fear of scorching the food or drying it too rapidly. Keep a thermometer inside and check it occasionally during the drying process. If it is an excessively warm day you may want to put in a smaller light bulb.

One of my neighbours has used this type of dryer for a number of years. He especially likes the ease with which it can be kept clean—he just washes it down, inside and out, with his garden hose.

In spite of the holes drilled in the bottom, he found moisture collected inside the refrigerator and didn't escape fast enough. He solved this problem by peeling off the rubber insulation strip around the door. This let the condensed moisture escape by giving it a place to run off and by improving ventilation, thus facilitating evaporation. (This is a problem only in the early stages of the drying, of course.)

The following drawings show how the refrigerator was adapted into a very successful dryer. If you are unable to procure a similar model with the fan already built in, you can set a $\frac{1}{16}$ horsepower fan in the bottom. Cut out a section big enough to accommodate it.

This same man also adapted a smaller refrigerator and made it into a smoker for meat and fish. He uses smoke flavoured curing salt and then freezes the meat, as it doesn't keep as well otherwise, but of course it has a unique flavour and takes up much less freezer space than if it were fresh.

Thousands of discarded refrigerators clutter up garages, sheds and scrap yards. For only a few pounds they can be converted into food dryers that will supply a family with good eating for years to come.

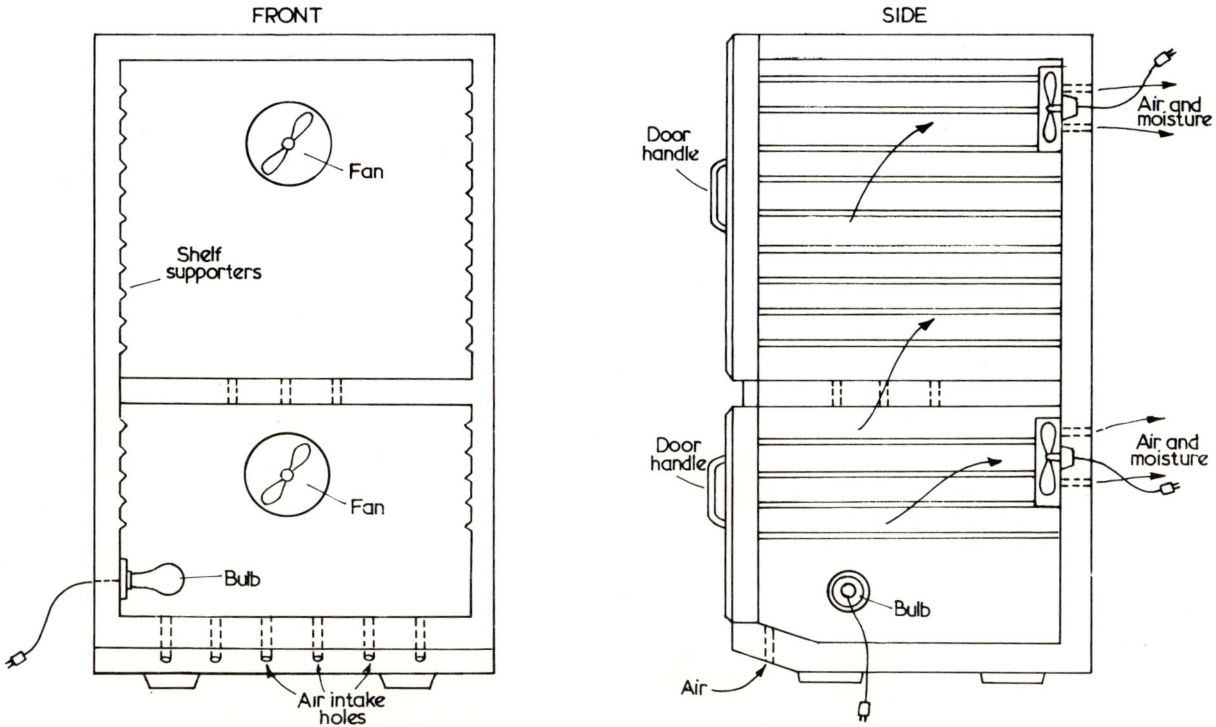

Fig 18 Dryer converted from a refrigerator

Oven Drying

If you are drying only small amounts of food, oven drying may be worth the effort. However, because there is almost no air circulation in an oven, it will require constant attention.

You can construct trays to fit your own oven, of fibreglass screening cut to fit the oven racks. Be sure the grill units of your oven, whether gas or electric, do not get turned on accidentally. Set your heat regulator at 65°C (150°F) for the first part of the drying period, then cut it down to 60°C (140°F) when the food appears to be half done.

Keep electric ranges open about 2·5cm (1in). Gas ovens should be propped open to about 20cm (8in) for the entire drying period. This will help to control the temperature and allow the condensed moisture to escape.

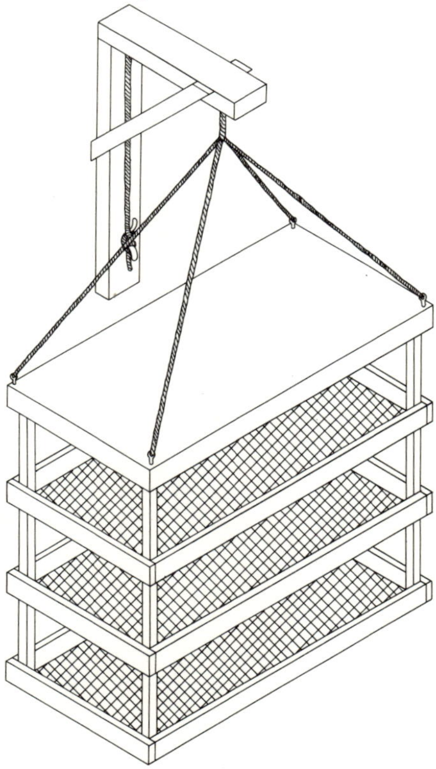

Fig 19 Hanging dryer

Check the temperature with a thermometer if you have any doubts about the accuracy of your oven control. Stir the food occasionally for even drying. And be sure there are 7·5cm (3in) of space between each shelf.

It is a good idea to keep the kitchen well ventilated during the 4 to 12 hours it will take to dry an oven-load of food. If you are satisfied the food has dried evenly, you can store it after it has cooled, with no further treatment. For the last 10 minutes of drying raise the temperature to 79°C (175°F) for pasteurization.

Over the Register Dryers

The same principle applies to this type of dryer as to all the others. It is useful for drying small amounts of food, and the important thing is to make sure the food had dried evenly and no spoilage has developed.

We found the best use for this type of dryer was for drying nuts. A neighbour takes regular dryer trays and places them on the overhead ventilator chutes leading from her forced air furnace in the basement to the rest of the house. This is another good way to use the heat that is already warming your home. It is best to use this only for nuts still in their shells, however, because of the dust and dirt found in basements.

Objections have been raised to this drying method because of the possibility of fumes coming from the furnace. Unless the furnace is very old this is unlikely to cause much of a problem. In any case if fumes are a possibility you should have your furnace checked, whether or not you intend to dry food over the registers.

Hanging Dryers

These can be used over almost any heat source. We have friends who live in the country and use a wood stove for much of their cooking and heating. If you keep a stove going most of the day and night, a range top dryer might prove useful. A simple way to build one would be to make the trays first, then attach them by braces to form one stacked unit, at least 15cm (6in) apart. Suspend this from the ceiling. Be sure you do not cook any strong-smelling foods on the stove while drying food, as the odours will be absorbed into the dried food. The dryer should be made so it can be pulled out of the way easily when not in use.

A commercially produced dryer of this type for use on top of your cooker utilizing its heat source is the Homestead Stove Top Dryer available by mail order from the seed catalogue of Thompson & Morgan (Ipswich) Ltd, Crane Hall, London Road, Ipswich, Suffolk. Constructed of aluminium, its outer dimensions are approximately 45×40×35cm (18″×16″×14″), and it weighs only 5kg (12lb). It contains five fibreglass screen trays and will dry an average load of 5kg (12lb) of food. The dryer comes with a booklet and costs approximately £50.

Trays

I have mentioned dryer racks, or trays as they are usually called, but feel they should be discussed more fully. A tray is actually a sliding shelf, made so that air can easily pass through. The ideal shelf does not exist, but if it did it would have the following properties: no air resistance; rigidity, so that it would support heavy fruit without sagging down onto the next tray; no chemical reactions with the food; easy to clean; non-stick surface; and it would be lightweight and durable.

Fibreglass screening works quite well except that it sags and, with heavy fruit, this causes problems. I like to use 6mm (¼″) galvanized hardware cloth, since its rigidity adds to the strength of the frame. Remember, however, to make allowances for the space it will take up, so that the rack can still slide easily.

With fibreglass there is no reaction of any kind to the food, but, before its initial use, it should be thoroughly washed to rid it of a substance sometimes left after extrusion from manufacturing moulds.

In Great Britain, the Netlon Company has an excellent small mesh made of non-toxic plastic which serves quite well, although it may need an extra strip of wood under the middle section of the tray for added strength.

Wooden slats are also sometimes used. While this makes the tray more rigid it tends to cut down somewhat on air flow and also makes the rack a little heavier to handle.

If you do decide to use wood slats select very smooth wood as food will tend to stick to it. Wood also has a slight tendency to absorb odours and you may find the slats harder to clean than the screening.

In fastening either fibreglass screening or hardware cloth to the trays, I use a heavy duty staple gun. This eases the job and takes some of the strain off the tray frame. If you do not have a staple gun, and can't hire one, drive the staples in with the single edge of the frame set on a solid surface. You can also control the sagging a bit by putting in cross braces, then fasten the screening to them.

6 Drying Step-by-step

Selection of Food

While drying is a marvellous process, you cannot expect it to make a silk purse out of a sow's ear. You must start with good food to end up with good food.

Select fresh, ripe fruit and vegetables at their peak. Remove any blemishes or spoiled parts. One off-flavour morsel can spoil an entire tray. Keep knives sharp. This will speed up your work and avoid bruising. Be sure all work surfaces, utensils and trays are clean. Just as in canning and freezing, you must work fast from preparation to processing to avoid loss of flavour.

The beginner may want to start with such universal favourites as apples, apricots, peaches or plums. We have also dried cherries, bananas and pineapples with good results, and pears were a surprisingly good treat when dried. Because of their juiciness, berries tend to lose something in the drying process, but can nevertheless be dried with some success.

A friend of ours is experimenting now with drying blueberries from his garden to use in place of currants, which are now in short supply and expensive.

Vegetables that turn out well are corn, onions, celery, tomatoes, carrots, cabbage and the pumpkin family to name just a few. See Chapter 1 for an extensive list of foods to dry.

Flavour and Colour

Vegetables as a rule should be blanched (suspended over boiling water) in order to stop or slow down enzyme action, which can rob them of much flavour and food value. The length of blanching time will vary, and is given in Chapter 7. This step is not necessary with fruit, but if the fruit is dried whole, like cherries or prunes, sometimes a boiling water bath is recommended to check or crack the skin so that moisture can freely escape from inside the fruit.

This is accomplished in commercial processing of prunes by first dipping the fruit in a hot lye solution. This is dangerous for the home food dryer, because lye (caustic soda) can cause dangerous and painful burns if spilled on the skin. It also destroys some of the B vitamins and Vitamin C. For home drying we suggest you first remove the stones and place the fruit with the skin side down on the trays. This speeds up the drying process by exposing more cut surface to the air, and incidentally prevents the fruit from sticking to the tray, a problem if it is placed cut side down.

Place fruit on the tray in the amount of (¾–1kg) (2–2½lb) per square foot. Weigh the fruit the first few times to help you estimate the capacity of your dryer. This will make it easier to plan how much fruit to prepare at a time.

Arrangement of Food on Trays

Be sure the air can always move freely around the food, with no stacking or overlapping. Drying time will depend to a great extent on the thickness of the particles. A quartered apple, for example, will dry much faster than a halved apple. And apple slices 3mm (⅛″) thick will dry faster still. In drying, the moisture must travel from the very centre of each piece of food to the outside air. The thicker the piece the longer the drying time.

Don't be tempted to overload the trays. You will only end up with uneven drying and sometimes mouldy food will result. When the food has shrunk, usually when you are about halfway through the drying process, you can then add to the load, putting the nearly dried food on the bottom and the fresh food on the top trays. But it must be stressed, piling the food impedes air circulation and the food just will not dry satisfactorily.

Uniformity of size also makes a great difference since large pieces may still be moist while smaller pieces may be overdried and hard.

While you are preparing the food, begin to warm up the dryer. Do not open any of the vents at this time if you are using a box dryer. Recirculate the heating air until the dryer has reached the desired temperature. This is where a thermostat can be really helpful.

Discolouration of Food

Darkening or discolouration of certain foods, due to the cut surfaces being exposed to air, is distasteful to some people although this discolouration has little effect on food value. Several processes are used to prevent this. Commercially dried fruits are sulphured, not only to help maintain their colour but also, according to some authorities, to cut down on insect infestation, and according to a leading British authority (*Manual of Nutrition*, Ministry of Agriculture, Fisheries and Food, p 124) sulphur acts as a preservative by inhibiting the growth of micro-organisms.

A US authority (Utah State University Extension Service) says:

> Sulphuring has the advantage of giving a better quality product . . . will maintain colour, flavour, vitamins A and C, and will discourage insect infestation during drying. Exposure to moderate (recommended amounts) of sulphur or sulphur fumes is beneficial to the product and non-injurious to health. The heat of drying and subsequent cooking dissipates the sulphur.

The British *Manual of Nutrition* also advises that fruit preserved with sulphur should be thoroughly boiled to be sure the preservative has been removed, before it can be eaten. Although this of course refers to the use of sulphur dioxide tablets, and not to the smoking method, it gives an indication of how the ingestion of sulphur by humans is regarded.

There are other experts who point out that sulphur is a chemical found in the body and is therefore in a sense natural. We mention these points of view so as to help readers make informed decisions before definitely committing themselves, for or against the use of sulphur.

We personally tried sulphuring, and found it left our throats dry as if we were breathing the fires of hell for hours afterwards, so we don't use it. If the drying process is thorough and the food is pasteurized afterwards, then cooled and stored as recommended in this book, you shouldn't need sulphur.

Directions for sulphuring can be obtained from most Extension Services in the US but we include a recipe below from a family with a long and successful record of drying food.

Sulphuring Fruit

Purchase sodium sulphate from a chemist. Make a solution of 1 tsp to 4½l (1 imperial gallon) (slightly less for a US gallon) of water. Soak the fruit in this for 10 minutes, drain and place in dryer. The fruit will come out fairly free of discolouration which is why this solution is especially recommended for apples, peaches and pears, all of which discolour rapidly when exposed to air.

As we mentioned earlier, *never* put sulphured food on trays made from galvanized wire (also known in the US as hardware cloth). Harmful chemical reactions will result.

Somewhat easier to use is 'potato white' as it is known in the restaurant trade. It is available from catering supply houses. A combination of sodium bicarbonate and sodium bisulphite, it is used to prevent discolouration of prepared food, especially potatoes, and has been used with good results by friends who dry large quantities of food every year.

Ascorbic Acid Bath

If you prefer not to use sulphur, you might try the ascorbic acid bath. In Great Britain you can purchase a powder called 'Linus'. Some chemists do carry this but it is also readily available in health food shops. (In the US this powder is obtainable from drug stores and health food stores.)

Make a solution of 2½ tsp to each 225g (8oz) cup of cold water for apples; 1 tsp per cup for pears, apricots, peaches and also potatoes. Sprinkle this on after the food has been cut and prepared for the trays. Or, dip fruit in solution in shallow bowl, being careful to cover all cut surfaces and avoid bruising.

You might also try mixing 1 tbsp of citric acid, *or* 500 milligrammes of vitamin C, obtained either by crushing the tablets in water or purchasing the vitamin in liquid form. Mix either of the above amounts in 900g (32oz) water. This will make a good, safe antioxidant.

Salt Water Bath

This is sometimes used but we do not recommend it as water-soluble vitamins like B and C, minerals, and sugars are lost in the process. It also adds more water which of course necessitates a longer drying time. It is done by soaking the fruit in a solution of 4 tbsp salt to 4½l (1 gallon (imperial)) of water. Some people add a cup of sugar to this solution. Drain thoroughly.

To Peel or Not to Peel

Whether or not to peel fruit and vegetables depends partly on individual preference and partly on the type of food. Of course you will always peel aubergine (eggplant) but there is a difference of opinion about peeling fruit, for example, or even potatoes. Some people automatically peel everything because they have always done so. Others have to peel some foods because they cannot digest the tough skins, or because they dislike the taste and texture.

I find the peel one of the most tasty parts of the fruit, adding welcome tartness to the natural sweetness. I leave the peel on as long as it doesn't make the food taste too tough or too bitter. Nutritionally it is far better to leave it on because of the valuable minerals that lie just beneath the skin of most fruits and vegetables.

I did find that it was better to peel marrows, but young courgettes (zucchini) could be dried and cooked with the skins left on. I never peel apples, carrots, potatoes, tomatoes, or cucumbers, and dried peaches and pears are delicious with the skins intact.

Unfortunately much produce now is sprayed, although conscientious efforts are made to lower the risk of poisonous substances reaching the public through food. This risk must be balanced against the need to guarantee a crop for the farmer and food for the population.

Rather than peeling food that has been sprayed, you can wash it first in a very dilute solution of vinegar, then prepare it for drying. Organically grown produce doesn't need this, and we use it whenever we can get it, partly for this reason and partly because we think the flavour is superior to food grown in chemically fertilized soil.

Into the Dryer

When the food is first placed in the dryer (if you are using the box type) open both intake and outlet doors. During this stage, the air should be kept at about 60 to 80 per cent humidity. I find a happy medium between these two levels has been reached when I look into the dryer and my spectacles fog! This shows that the moisture is getting out and the air is pretty well saturated as it leaves.

During the initial stage of dehydration, temperature is not too critical. Evaporation of the moisture from the food will keep the dryer cool during the first few hours, and may even cause the temperature to drop below 37°C (100°F). This adds only slightly to the total drying time, and you will end up with a better product. Some authorities recommend higher temperatures, up to 79°C (175°F), but we are inclined to believe this may be partly because most research in food drying is based on commercial methods, where higher temperatures are used.

We experimented with these higher temperatures in our box dryer but are satisfied that the lower temperatures produce better-tasting food with excellent keeping properties. We opened a packet of fruit recently that had been stored for four years, and it was still fragrant and full of flavour. It was dried at the lower temperature and for this reason we recommend it.

As the food is being heated and dried, there will be less water evaporating therefore the temperature inside the dryer will tend to rise. If you have not put a thermostat on your dryer, you must watch it very closely during the final stages. It is at this time that most scorching, with a consequent loss in flavour and quality, occurs. The same amount of heat which barely kept the dehydrator at about 30°C (in the 90s Farenheit), is now liable to shoot the temperature up to 65°C (150°F) or more.

We usually let each batch go on to completion, then pack it away before starting on the next one. However, if you have prepared more than will fit into the dryer, and have enough left over for one or two more trays, wait until the food in the dryer has decreased in bulk by half, as it can then be fitted onto fewer trays. This will leave room for the rest of the prepared food. This is a good way to add leafy foods to a dryer load, as they dry much more rapidly than thick-stemmed foods. (See celery, Chaper 7.)

Always place the fresh food on a tray by itself. Because of its heavier weight and extra moisture content, fruit especially must be kept separated from any partly-dried food.

Remember this important rule: *never combine different types of foods, especially those with different drying times, on the same tray.* They can all go into the dryer at the same time, but leave each individual tray in the dryer until the food on it tests dry.

When adding a tray of fresh food after one batch has already been started, place it on the top shelf, or whichever shelf in your particular dryer dries the fastest.

Rotation of Trays

When the dryer is kept between 32°C to 43°C (90–110°F) tray rotation is usually unnecessary. As long as your dryer has complete air movement, one shelf is much like the next.

On the other hand, food benefits from a change in position to ensure regularity of heating. Since heat and moisture both tend to rise, rotating the trays will help to eliminate uneven distribution of heat and unwanted concentration of moisture.

Some authorities say food dries fastest around the edge of the trays, others that the centre of the trays is the most efficient spot. We have used a number of different dryers and found they vary from one to another. As you continue to dry food, you will become familiar with the properties of your own model, and will soon discover whether or not you have to rotate the trays and how frequently this may be necessary.

We do recommend stirring the contents of the trays occasionally after the drying has been going on a few hours. It also helps to reverse the trays, end for end, at least once, all with the aim of promoting even drying. All these instructions emphasize the advantage of using a thermostat which, with the right design of dryer, should make tray rotation almost unnecessary. We know there are some who advise turning off the dryer at night, but with the low temperatures we recommend and with a thermostat this should not be necessary.

If all this sounds onerous and too much to bother with, let me assure you it isn't. In fact, I find I can't resist looking in the dryer to see how the food is coming along. And usually the lovely smell of food drying will remind you to check on the trays. After a few trial runs, drying food should prove an easy, enjoyable process which brings rich rewards in good eating and economy.

Problems in Drying

Mould Food did not dry thoroughly. Always make a test with several pieces of food from each tray before removing from the dryer. Fruit should be leathery but not stiff. Vegetables should be brittle. Herbs should crumble to the touch. See chart in Chapter 7 for individual food tests. For final test, let food cool a few minutes as it will always be more limp while warm.

Rot See Mould. Storage must be secure from light, air, moisture and rodent and insect infestation. Check containers for tight seal. Do not store in warm area, nor near highly scented substances if stored in plastic containers.

Weevils Pasteurization, if directions are followed, will effectively kill weevils and eggs. Check instructions, and if weevils have not become too thick, repasteurize. You will find little dried-up worms on the bottom of the pans afterwards. If this is too offensive, discard the food. However, it is only fair to point out that most food is sold under health regulations which specify allowable levels of weevils and other forms of filth, perhaps more than the amounts you may find after you have repasteurized your food.

Too dry Dryer temperature was too high for too long a period. It is better not to be impatient and to allow the few extra hours it may take to finish the food properly, than to turn up the thermostat and risk scorching and hardening the food.

Too hard Initial temperature was too high.

Poor flavour Food was not good initially, temperatures may have been too high. Food may have been soaked in water beforehand. Sun drying sometimes leaches out flavour. Good food, carefully prepared and properly dried, will have concentrated natural flavour.

The following list is a guide to how much dried food you can expect from the fresh product.

Fruit	**Fresh**	**Dried**
Apples	2½–3½kg (6–8lb)	½kg (1lb)
Apricots	2½kg (6lb)	½kg (1lb)
Cherries	1½–2¼kg (4–5lb)	½kg (1lb) (should be stoned first)
Figs	2½–3½kg (6–8lb)	½kg (1lb)
Peaches	2½–3½kg (6–8lb)	½kg (1lb)
Pears	2¼kg (5lb)	½kg (1lb)
Plums	2¼kg (5lb)	½kg (1lb)

Vegetables	**Fresh**	**Dried**
Asparagus	1½kg (4lb)	½kg (1lb)
Carrots	3½kg (8lb)	½kg (1lb)
Celery	4½kg (10lb)	½kg (1lb)
Mushrooms	3½kg (8lb)	½kg (1lb)
Onions	3½–4½kg (8–10lb)	½kg (1lb)
Potatoes	2¼kg (5lb)	½kg (1lb)
Green Beans	3½kg (8lb)	½kg (1lb)
Tomatoes	4½–5½kg (10–12lb)	½kg (1lb)

These are of course approximate, but they will help you plan a dryer load so you won't prepare more food than you can process at a time.

7 Food Preparation

Outline of steps for drying from food preparation to final storage: select firm, fresh, ripe food; start up dryer to preheat. If sun drying, have trays ready so food can be placed immediately in dryer; assemble the following items:

- sharp stainless steel knives
- apple corer, peeler
- bowls to hold dipping solutions and prepared food
- deep saucepan with colander or wire basket for blanching
- plastic wrap for juicy foods
- clean towelling and sheeting for draining moisture from food after blanching.

You will also need one of the following: heavy nylon netting, butter muslin, cheesecloth, or old curtains or sheets to put under juicy foods.

Before you start out make sure you have oil to apply to trays to prevent sticking, storage containers (not needed until the food has finished drying, but it is better to have them ready before you begin); baking sheets for pasteurization in the oven after drying; plenty of old newspapers to spread under the table to catch excess juice, dropped food, etc.

Charity shops, jumble and rummage sales or garage sales are all good sources of old curtains, towelling and sheeting, to use for draining food after blanching. Old net curtains are good for spreading on trays when drying sticky foods. Use a few times, washing after each use, and when they fall apart you can discard them without feeling wasteful.

First wash all food thoroughly. Then sort through for spoiled or blemished pieces and discard. Peel, slice, core, shred, pit, etc as required. Dip fruit in ascorbic acid anti-oxidant solution; or sulphur, blanch or steam as required. Blot excess moisture. Place on dryer trays and into dryer. Sweet fruit, vegetable marrows and tomatoes may stick to the trays. To minimize the likelihood of this, brush surface of trays thinly with vegetable oil, or spread cheesecloth or other light material on trays before placing the food on them. This material can usually be washed and reused a few times before being discarded. Always wash new cheesecloth before using, as it contains a dressing which could scorch from the heat of the dryer.

Check trays during the drying period for evenness: rotate top and bottom trays if necessary to speed drying time or if food is not in uniform pieces. Condition if necessary. (See page 73.) Place on baking sheets and pasteurize in oven. Cool and store as directed in Chapter 3.

Always test to make sure food is dry enough. See chart at end of Chapter 8 for dry test information on individual foods.

Size of Prepared Food Pieces

Food for the dryer should be cut in small pieces, as uniform as possible, for more even drying results. Sharp stainless steel blades not only make the work go faster, but reduce bruising and discolouration of food.

Serving size is the proper criterion. This lessens the amount of preparation time required when the food is reconstituted. And chopping big chunks of apple, for pie or cobbler, is a tiresome and time-consuming chore. Most food, especially fruit, is more easily diced, chopped or sliced when fresh than when dried.

Naturally moisture content of the food has much to do with determining the size. Because water travels easily to the surface of woody vegetables for evaporation they can be cut in larger pieces. Older carrots, for example, give up their moisture much more easily and so can be dried in larger pieces than young, tender carrots.

When cutting fruits and vegetables, it is better to cut across the fibre than with it. This will make the rehydrated food less tough and also aid in faster drying. The prime consideration is always aid in the evaporation of water from the food. Vegetables dried commercially are generally cut in strips 6 × 6mm ($\frac{1}{4}'' \times \frac{1}{4}''$).

Blanching

Enzymes are necessary for life, because they aid essential

chemical reactions in the body. However, when we dehydrate food we do not want to encourage these reactions, because they cause food to turn brown and become tough, resulting in a product of lower quality.

So unless enzyme action is slowed down or halted, the food will continue to progress on its normal cycle which of course ends up in decomposition. When we dry food, we try to inhibit enzyme reactions, and in vegetables this is best accomplished by blanching.

Steam Blanching v Water Blanching We know people who never blanch anything, some who swear by blanching, and then there are arguments as to whether steam blanching or water blanching is best. After much experimentation and research, we cast our vote for steam blanching. It does not leach out minerals and vitamins as water blanching does, although it does take up to one-third more time.

It is true it may not be quite as reliable. A plunge into boiling water is a more violent method, so with water blanching you are pretty sure you are getting the heat to every portion of the food. With steam blanching you have to test the centres of the food pieces to be sure they have been thoroughly heated (but *not* cooked).

If you do use water blanching, we strongly recommend you use the blanching water for soup or other cooking, in order to save the precious minerals left after blanching.

Blanching not only halts enzyme action, but breaks down some of the cell walls of the food, making it easier for water to evaporate. If your time is limited, however, it is permissible to omit this step rather than have the food go to waste. The penalty will probably be a slightly lower quality in the finished product.

Most vegetables dried without blanching tend to be a bit woody in texture and flavour. Exceptions to this are tomatoes, onions and mushrooms.

Steam blanching After preparing the food according to individual instructions, suspend a couple of inches above a vessel of rapidly boiling water for the time recommended. Keep covered tightly to contain steam.

Water blanching Plunge the food in boiling water, using a wire basket to make it easier to remove, cover tightly, and leave in only until translucent. Do not cook. You can use a piece of thin cloth, like cheesecloth, to tie the food up and plunge this in the pot. We have to disagree with some authorities who recommend long blanching times. We have done many experiments on various foods and found most recommended times were longer than needed.

Do not use your pressure cooker for blanching unless you leave the petcock open and the cover just loosely placed on top. In other words, use it as another pot, not as a pressure cooker. The purpose of blanching is to break down some of the hard cellular material and halt enzyme action, not to cook the food.

Be sure the food is loosely arranged so all pieces are penetrated by the heat sufficiently. It is a wise precaution to test the foods in the centre of the blanching container when half the recommended time has elapsed to be sure that all parts are being reached. When the entire time has passed, the food should be heated through and wilted in appearance.

Usually the thicker portions of a food, like the stems of broccoli or the thick central part of spinach or chard, require more time than the leaves. Process the thicker parts first, then remove and add the rest.

When blanching is completed, remove food from boiling water and plunge at once into cold, or even iced water if possible. Allow to drain, pat off excess moisture, then start on the next batch.

Be sure to bring the water to a rolling boil before dipping the new batch. Also, it's better not to use the same water for different foods as the flavour and odour will be transferred.

Blanching should not be used for substances with a great deal of flavour such as garlic and onions, as the volatile oils in these foods will be diminished by blanching.

Summary of Blanching It should be used mainly for vegetables, although some experts recommend it for fruit also. Water must be boiling during the entire process. Vegetables are finished when they have a translucent appearance.

Use a basket, strainer or colander to dip food. Reuse the same water for the same food, for as long as possible. Since there is always some loss of nutrients when food is subjected to water or heat, the boiling water will eventually become saturated and then not as much of the mineral content of the food will leach out.

Advantages of Blanching Prevents darkening and aids in retention of natural colour. Softens tough fibres so they will give up moisture more readily. Inactivates spoilage organisms. Expels some of the oxygen from the food, helping to prevent oxidation while drying. Cuts down on enzyme activity.

Conditioning

When the dried food is ready to be packaged and stored, you must be sure that all pieces have dried evenly. One moist piece can spoil an entire batch by allowing mould or rot to develop.

In commercial drying, food is conditioned by processing it in large quantities to 'cure' it, the damper pieces giving up moisture to the less damp. The same process can be used by the home dryer for the much smaller amounts to be dealt with.

If you have doubts about how evenly the food has dried, it is best to condition it according to the following directions. In general the larger the pieces, the more likely they will need conditioning.

Remove food from dryer when it meets drying test, and allow to cool. Pour into large containers such as washtubs, large bottling (canning) utensils, crocks, etc, which have been lined first with plastic sheeting, then with clean cotton sheets. Cover food with screening or tightly stretched nylon netting, and set in well-ventilated room, cool but not cold, and protected from insects, rodents and excessive moisture. Stir contents at least once a day, and continue to do so for ten days or until food particles seem uniformly dry.

Clint Seibert calls this 'seasoning' and has this comment to make:

> I like to season my food by placing in a brown paper sack at room temperature and leaving the top open, hanging it about head high in the room. For years I wondered why shredded

carrots moulded over the winter. Finally I left some to season for a month and it kept for a whole year.

We have dried many batches of food and found conditioning was unnecessary as long as we observed the following rules: never let the temperature go above 43°C (110°F); about halfway through the drying process, rotate trays top to bottom and end to end, if you are drying extra large pieces or extra juicy food.

Pasteurization

While conditioning should free food from the last vestiges of moisture, insect eggs many be acquired in the process. Therefore pasteurization should be the next step. Even if you do not find it necessary to condition the food, we believe pasteurization is essential. Since most food is dried during the warm summer months when there are always insects about, the chances are a stray bug or two may get a chance to lay some eggs on your nicely dried fruit or vegetables. Pasteurization makes sure they don't get a chance to develop into worms which will enjoy your food before you get a chance to.

Spread food loosely, about 2·5cm (1") in depth, on baking sheets and place in oven preheated to 80°C (175°F). Allow 15 minutes for vegetables, 20 minutes for fruit. Cool in an area protected from insects. Package and store immediately, or food may reabsorb moisture. Be sure the oven does not go above the recommended temperature, or the food will harden.

In following instructions for individual foods, it is understood that after drying all foods are: conditioned (optional), pasteurized, cooled, then packaged immediately.

Individual Instructions: Fruit

The length of drying time for fruit depends on: size of pieces; temperature of dryer; movement of air in dryer; juiciness of food.

Wherever discolouration of the food is important to you, we suggest the ascorbic acid bath (page 63).

Apples Late, firm varieties are best for drying. Wash, peel (if desired), core, and either: slice in 3mm ($\frac{1}{8}$") thick rings; cut into eighths; cut into quarters; grate or shred coarsely, as for apple sauce.

If grated, line trays with plastic or netting. To maintain colour (if desired but not necessary), dip in ascorbic acid solution. Or blanch first, 4–5 minutes. Blot excess moisture. Place in dryer.

Apples are the all-time versatile favourite for home drying. We believe you can never dry too many of them. They are perfect for apple sauce, pies, cobblers, crisps, cakes, sponges, and eating on their own. Try dried apples for TV snacks, lunchboxes, camping trips and a handy titbit to keep in the car or in your desk drawer. Also see **Leather.**

Apricots Peel or not as desired. Cut in halves, remove stones. Blanch 4 minutes (optional); blot excess moisture. Place fruit cut side up on trays. Larger fruit should be sliced or quartered. Dry, pasteurize, cool, store. Also see **Leather**.

Berries Dry whole, on trays lined with netting to prevent sticking, as with all juicy foods. Some authorities recommend slicing very large berries as they may mould in the centre before drying can be finished. They must be completely dry before storing. Also see **Leather.**

Cherries Stone, drain off excess moisture. As with berries, no further processing is needed before drying. Pie cherries make a good substitute for raisins, giving a bit more tartness than when grapes are used. Also see **Leather.**

Bananas Use only very ripe bananas, with brown flecks on the skin. These are often sold at reduced prices. Remove blemishes, slice in 6mm ($\frac{1}{4}$") pieces, or cut lengthwise in quarters. Very small bananas may be dried whole. No processing needed.

Bananas are one of the most delicious of all dried fruits. They may be eaten as they come out of the dryer, or powdered in the blender for flavouring breads, cakes, frostings, candies, milk-shakes, etc. Also see **Leather.**

Citrus Rinds Use only organically grown fruit, to be sure there is no poisonous residue. Dry without any processing. Best cut in narrow strips. Will make a valuable addition to your spice shelf. May be chopped, or powdered in the blender when dried.

Cranberries Dry when prices are lowest, otherwise they aren't really worth the trouble. Because of their tough skins we recommend a 30-second bath in boiling water (use a mesh basket, colander or large strainer), followed by a quick dip in very cold, preferably iced, water. Blot excess moisture. Dry until very hard, with no moisture when berry is cracked open. Pasteurize, store. Use throughout the year for cranberry sauce and cranberry-orange relish.

Figs Must be very ripe. Wash, dip quickly in boiling water to crack skins. Dry whole or split in half if fruit is very large. Because figs are dried in larger pieces than is usually recommended, they should be turned over halfway through the drying, which can take up to 48 hours or a bit longer. Be sure they are completely dry in the centres.

Grapes (also currants) Use seedless varieties only. Wash, dip in boiling water bath to crack skins. Be sure to remove all stems. Dry, pasteurize, store.

Guavas Peel, cut in 3mm ($\frac{1}{8}''$) thick pieces. If you dip these in the ascorbic acid solution, use a large mesh strainer or soup strainer to lift the slippery pieces out. Spread on plastic film-lined trays to dry.

Nectarines See Apricots.

Papayas See Guavas.

Peaches See Apricots.

Pears See Apricots.

One improvement we noticed was to cut pears into strips, rather than halving or quartering them. The strips are leathery when finished but somehow the flavour is much improved in this form.

Another interesting variation for dried pears: combine four parts orange juice with one part lemon juice, and honey to taste. Dip the prepared pears in this, and drain. Dry as usual. The pears will be a bit darker in colour but will have a superb flavour.

Plums and Prunes Wash, remove stones, dip in boiling water bath to crack skins. Optional blanch 15 minutes. Blot dry. Place cut side up on tray.

Plums dry much faster when pitted, are easier to eat right from the package and take up less room in storage. As with pie cherries, ground prunes (pitted, of course) make a good substitute for raisins. When dry, they will test a bit sticky. Also see **Leather.**

Pineapple Only very ripe fruit should be dried. Peel and core. (I also dry the core, although it is hard and not quite as fully flavoured as the rest of the fruit.) Dice or slice in 12mm (½″) thicknesses. Spread on waxed paper or plastic film because of the juiciness. Scrape the pulp left on the skin and spread on plastic-lined trays to dry. Pineapple takes several days to dry but it is one of the most delicious fruits I have tried, with a concentrated sweetness and flavour. Use it dry for eating on its own. Reconstitute for icings, cakes, milk shakes, or puddings, or if you make your own ice cream. Also see **Leather.**

Rhubarb Wash, slice 12mm (½″) thick, needs no further processing before drying. Do not use the leaves. They contain oxalic acid, which is poisonous if eaten in large quantities.

Leather

Basically made from a puree, leather is usually made from fruit and is often spiked with chopped nuts, finely chopped or powdered citrus rinds, and seasonings like cinnamon, nutmeg, coriander, allspice, etc. It can be sweetened with honey, sugar, molasses, raisins, dates, or figs, and sometimes flavourings like vanilla, almond extract or lemon extract are used.

Fruit leather is unsurpassed as a healthful, delicious and

economical treat for both children and adults, especially when compared with commercially made sweets. Combinations such as apple with banana, or berry, or pineapple and banana, pears and pineapple, the list is endless.

Try blending a nice-tasting protein powder with one or more of these fruits for a highly nourishing snack food. My family's problem is that we usually find the leather is eaten before we can get it stored!

To make Leather Prepare fruit as for drying, including blanching if desired. Chop and place in blender, adding only the minimum amount of liquid according to your blender directions. Very juicy fruits, such as berries or plums, require no additional liquid.

Blend fruit, adding flavourings and sweetenings to taste. Do not add nuts, if used, until you are ready to place mixture in dryer, or they will be ground too finely in the blender. For thickening, if needed, add a pinch of gelatin, agar-agar or if you can find these at the health shops, flaxseed or chia seeds. If you use apples you may not need any thickening, because of their high pectin content. This makes them a favourite fruit to use in combination with others.

Consistency should be like a thick apple sauce, runny but definitely not a liquid. The runnier it is, the longer it will take to dry and the stickier it will be.

Be sure to test consistency before spreading on plastic-lined dryer trays. I usually let the mixture sit in the blender for about 5 minutes to let it thicken up. As for flavour, the finished leather should taste exactly as it does before drying.

You will save yourself trouble if you fasten the plastic wrap to the trays either with sellotape or with thumb tacks, as the weight of the leather sometimes drags the plastic down, covering the leather around the edges. This prevents that portion of the leather from drying. For this reason I suggest you do not spread the mixture to the edges of the trays. Leave a border of bare plastic. This will not only speed up evaporation, but make it easier to peel the leather off when it has dried. Dry at 43°C (110°F). After 24 hours, or when the top layer is dry, lift it off the tray and turn it over, peeling off the plastic so the wet side is face up. Continue drying until the leather has a slightly sticky consistency, with a translucent appearance, and a chewy texture.

In some dryers, where the air circulation may not be quite so efficient, you may find that you will want to use only half the surface of each tray, staggering the food in order to leave alternate halves of the trays empty to let the air get to all the food. In a well-designed dryer, however, this should not be necessary. I mention it, however, in case you are using a dryer with an inadequate fan, or in case of the heat source fluctuation.

When finished, peel the leather off the plastic, then store in cool, dark, airtight containers. Glass jars are very good for leather, but be sure they are kept in a dark place. Cut the leather into strips for snacks, and carry some in your car for emergency hunger pangs, or if you work in an office, keep some handy for a quick energy pickup.

Leather will keep for a month in airtight containers, and can be frozen for a year with good results. You can also make leather from dried fruit. I have even heard of people making leather from canned fruit, because they were so fond of it and couldn't get fruit at the time.

Apple can be combined with: apricots, bananas, berries, cherries, grapes, peaches, pears, prunes or plums, raisins, dates. Soak dates in hot water until soft before blending: even better if you chop them as well.

Experiment with the above combinations, and try any others which sound interesting. Remember, though, not to add dates until you have finished blending if you want a bit of crunchiness. Toasted sesame seeds or sunflower seeds are also good in leather, and make a nice crunchy contrast to the smooth sweetness of fruit.

Do not use berries alone, because of their extra juiciness. Combine them with apples, or any of the thickeners mentioned above. Strain out seedy berries after blending, then return the mixture to the blender for whatever sweetening or thickening may be required. Grapes, if made into leather, will also need both sweetening and thickening.

Individual Instructions: Vegetables

Beans (String, snap, yellow, runner, wax) Sort through for imperfect beans. Wash, remove strings and ends if necessary. Blanch 15–20 minutes, drain. Split large pods lengthwise to hasten drying time. Dry.

Beets (young only) Wash, trim tops but leave beetroot intact. This is to keep the juice from bleeding during blanching. Blanch 30–45 minutes, or until almost tender. Cool, trim off top and roots, and peel if desired. Cut into very thin strips, or slices 3mm ($\frac{1}{8}''$) thick, grate or cube. Drain, dry, pasteurize, package and store.

Broccoli Wash, and prepare as for serving. Cut stalks in quarters, then in half. Blanch until nearly tender and translucent. Do not overcook as broccoli then becomes mushy when reconstituted. Drain, dry.

Cabbage Discard outer leaves, remove core, shred or chop in very thin strips. Blanch 5 minutes, until wilted. Pat dry, spread on trays. Dry, pasteurize, package, store. Cabbage is a vegetable which shrinks considerably and you can add fresh food to the half-dry as space becomes available in the dryer.

Carrots (young tender carrots preferably) Wash, trim, then either shred, or slice very thin, or cut in very thin strips. Blanch about 10 minutes, until almost tender. Drain, dry.

Celery Celery is not often dried commercially because its high initial cost as a fresh vegetable makes the dried product exorbitantly expensive. If your garden has some surplus celery, however, or if you manage to buy a crate cheap at the market, by all means dry some.

It is better to dry celery in two stages. Fill the trays with the stalks first. Then when they are almost dry, they will have shrunk enough so there is room to add the leaves. Place the leaves on separate trays. Both stalks and leaves should finish drying at the same time.

Stalks Wash thoroughly, discard any discoloured parts, especially the fibrous outer stalks. Cut larger stalks in half lengthwise; then dice all stalks about 6mm (¼″) thick. Blanch 4 minutes; drain, place in dryer.
Leaves Need only be washed and blotted dry before being placed on trays. Pasteurize, cool and store separately from the stalks.

Courgettes (zucchini) Drying this prolific vegetable was a delightful surprise. They taste a bit like mushrooms and can be eaten as they come out of the dryer as chips for party dips. Cooked, they are great in casserole dishes, or reconstituted, dipped in batter and fried.

Select only young, tender vegetables. Wash, cut in 6mm (¼″) slices, or dice about 12mm (½″) thick. Blanch 7 minutes, or until almost tender. Drain, dry. Pasteurize, package, store.

Garlic Not necessary to dry, as it can easily be purchased already dried. If you grow garlic, hang in cool, dark place and it should dry satisfactorily.

Marrows (large zucchini) Treat as for courgettes, but these vegetables must be peeled as the skin is very tough.

Mushrooms Wipe clean. (Avoid washing if possible, as they will lose flavour if you do.) Slice large mushrooms, but leave small ones as they are. Dip in ascorbic acid bath if you want to keep them a pale colour. We never blanch mushrooms, although some authorities suggest it. Drain, dry, pasteurize, package, store.

Mushrooms are a joy to have on hand. They can be powdered in the blender and used for flavouring soups, stews, etc, even dips. Dried mushrooms reconstitute so easily they can be used for quick last-minute dishes like omelettes, and I buy them up whenever I see them on sale and dry them. Some people string mushrooms, like apples, and hang them across the kitchen to dry.

Onions Remove outer layers. Slice very thinly, place in dryer. Other foods will pick up onion odour, so it is a good idea to line trays with plastic before drying onions, or dry with compatible foods such as soup vegetables etc. Wash trays after drying onions to remove odours.

Parsley (See Herbs).

Parsnips (See Potatoes).

Peas (young, tender only) Shell, blanch, usually only 3 minutes, rinse in very cold water, drain, and dry. Stir occasionally during first few hours, to be sure of adequate air circulation. Peas must test 'pebble dry', and should shatter when hit with a hammer.

Snow peas, or Chinese pod peas may also be dried, after similar pea-drying treatment. Pods should be leathery-tough when finished.

Peppers and Pimentos Wash, remove seeds and strings, cut into very thin strips, or dice thinly. Blanch 10 minutes, drain, dry.

Potatoes Scrub, peel if desired, cut into thin slices or shoestring slivers, or shred for later use in potato pancakes or little cakes. Blanch 3–5 minutes, drain, dry.

We did a lot of experimenting with potatoes, and found they turned out quite well with this very short blanching time. We also discovered that it is not necessary to use an anti-oxidant on them if you don't care too much about the slight discolouration that does occur. Dried potatoes really turned out to be a wonderful addition to our larder, a convenience food and an excellent standby when a quick meal was needed.

Pumpkin or Winter Squash Chop open, cut into 2·5cm (1″) strips, remove fibre, rind and seeds. Save the seeds and dry them separately. They make a wonderful snack and are rich in minerals and vitamins. Cut strips 3mm ($\frac{1}{8}$″) thick, blanch until tender but not cooked, 10–15 minutes. Drain, dry, pasteurize, pack, store.

An alternate method: peel, remove seeds and fibre, cut into manageable chunks and grate coarsely. Blanch 5 minutes, until tender. Pat dry, and place on trays lined with cheesecloth.

Spinach (Also other greens such as mustard, dandelion, beet tops, kale, chard, collards) Wash as soon as possible after picking. Blanch until wilted only, about 4 minutes. Chop the large coarse leaves crosswise. Drain, and spread only one layer deep on trays. After a few hours they will have shrunk enough so you can then put more food in the dryer on separate trays. Like herbs, greens should be crumbly when finished.

Tomatoes One of the surprises of drying, tomatoes retain their flavour, colour and aroma and can be made into very useful leather.

Select red, ripe, blemish-free tomatoes, dip in boiling water if you wish to remove the skins, but this is not necessary. Cool before peeling. Slice very thin or quarter. Cut smaller varieties in half. Blanch 2–3 minutes if desired, drain. If tomatoes are very juicy line trays with plastic. Dry.

Tomato leather Treat as above. Before placing in dryer, however, run through blender or food mill, then spread on trays lined with plastic film wrap. You may want to strain if you don't get the tomatoes completely puréed. When dry, leather will be thin and crackly and can be used in almost any recipe calling for cooked tomatoes or tomato flavouring. Very good if powdered, then used as needed for sauces, or tomato pastes.

Turnips (See Potatoes).

Zucchini (See Courgettes and marrows).

Herbs

The cook's best friend—versatile, dependable herbs contain volatile oils, so no herb should ever be dried above 38°C (100°F). Do not expose to sunlight after picking as this takes away their colour.

It is best to pick herbs in the morning, on a bright, sunny day, when they contain less moisture. Use only young, tender plants and select the leaves from the top part of the stem. Rinse off dust, insects, cobwebs, etc, and let dry on old sheeting for an hour or two away from the sun.

Before placing in dryer, remove tough stems and stalks. Or wait until after drying, when it is a simple matter to strip the dry leaves. If you pick wild plants be absolutely certain of your identification. This is even more important with mushrooms. Herb experts advise that plants be gathered far away from motorways as plants can accumulate poisons from the fumes of passing cars.

If you do not have a dryer you can dry herbs by tying them in bunches, then hanging them in large paper bags on a screened-in back porch to dry. If you spread them out to dry on a table, weight the paper bags down with stones or bricks to keep them from blowing away.

Herbs generally dry in 24 hours, and should be crumbly. Store in airtight jars or cans away from light. Some people like to powder them before storing.

Meat, Fowl, Fish and Game

Meat Dried meat ranks low on the priority list of today's families unless they include camping and long hikes into the country in their activities. Freezers have revolutionized our meat-buying habits, and with the increase in meat prices since the beginning of the 1970s, surplus meat is not much of a problem for most of us.

However there are times when, either through one's own hunting success, or access to farm animals, a large quantity of meat suddenly becomes available. Perhaps freezer space is not available. Or a camping trip is planned and dried meat would add variety to the menus.

Man has successfully dried meat, fish and fowl since before recorded history. Just as today, our ancestors wanted food for travelling which would be light in weight, sustaining and not subject to deterioration, and which could be eaten 'out of hand', since cooking was not always possible.

There is some controversy and disagreement among home economists about the advisability of home-drying meat. Yet so

Apples cut in eighths and in rings ready to be dried

Fresh apples and dried apples

Fresh onions and dried onions

Fresh pears and dried pears

many people have done it for so many thousands of years, we decided to include a section on it because, with modern methods, and our greater knowledge of the principles of food preservation, such food can be dried safely if the instructions are carefully followed.

It is important to remember that animal flesh is an almost perfect medium for bacteria to multiply. Therefore, care must always be taken in the processing of such foods.

This same danger, however, exists in all types of food preservation. Improper freezing can lead to spoilage, as well as failure to sterilize jars and lids properly when bottling.

The following instructions should give you excellent results. If you are not quite sure about your dried meat, store it in your freezer until you take off on that camping trip. You will still have the benefit of much lighter food to carry, which will take up less space, and probably cost less than tinned meat or fish.

Use only lean muscle meat, such as flank steaks, and orgau meats. Remove all fat and any skin; cut across the grain if you prefer tender meat, with the grain for a more chewy texture. Cut in very thin strips, no more than 6mm ($\frac{1}{4}''$) thick. Season with salt, seasoned salt, soy sauce, smoke flavourings, tabasco, pepper or poultry seasoning. Strips must be very dry when finished but be sure they have cooled before you test them. They will be a bit soft while still warm. The finished piece will look shrivelled and be quite dark in colour.

Store in airtight containers, away from light, and preferably below 4°C (40°F), or in freezer. Drying time will be from 6–15 hours.

Fish Remove skin from any fatty fish. Small varieties like smelt need not be skinned. Bone larger fish. Cut fish in small pieces, 12mm ($\frac{1}{2}''$) thick. Remember that as in all drying, the smaller and more even the particles, the faster and more evenly the food will dry.

For smaller fish it is not necessary to remove skin and bones. Cut fish in half and spread flat. Dry very hard, store as for meat.

Fowl Fowl may be dried either with or without pre-cooking.

Remove all fat, including skin. Cut into very thin slices, especially the giblets. Season as for meat. Poultry seasoning and salt are an especially good mixture. Finished pieces must be hard with no sign of moisture. Store as for meat.

Reconstituted meats are good for flavouring soups, stews and casseroles, and may be added cold to salads. In all cases, first reconstitute by soaking in water. If used immediately, or refrigerated, the water can be useful for cooking.

We include this suggestion from an experienced cook, mother of eight, and long-time advocate of food drying. Like many women who live on farms and have a surplus of eggs, she freezes them, but after a time storage can become a real problem. The following method offers a good solution.

Thaw frozen eggs, one packet at a time, by placing in refrigerator until just frosty but not completely solid. Put in blender to make particles more uniform, chopping into small pieces first. Spread very thinly on baking sheets which have been sprayed with an anti-stick preparation (in the US something like Pam will do). Place in your dehydrator until crumbly and dry. Scrape into plastic containers and return to freezer. They will take up much less space and cook up very well for scrambled eggs, casseroles and for baking. Eggs must be kept frozen to avoid rancidity.

I used eggs for three years after freezing them in this manner and they weren't bad, even when scrambled for breakfast, after all that time. Incidentally, to freeze eggs initially, all you need do is to break them into a bowl (usually I do two at a time, because it is easier to measure for later use), and beat with a fork just enough to break the yolks. Do not beat thoroughly. Add just a few grains of salt or sugar, then freeze in ice cube trays. Do not forget to label the eggs to show whether they are salted or sugared and use them accordingly afterwards (ie salted eggs in savoury omelettes, sugared eggs in cakes).

8 Food Drying in Flats

You do not need to live on, or even near, a farm to enjoy the benefits of home food drying. You need not even live in a good sized house, a flat (apartment) will do just as well. If you have a kitchen and a spare corner large enough to accommodate your dryer, you can dry foods successfully. Imagine the convenience of being able to store a year's supply of fruit and vegetables in a cupboard!

So you see, lack of space need not keep you from drying all the food you need to last from one growing season to the next. In fact, one of the great advantages of this method is that it requires far less space than either home bottling or freezing. In most homes and almost all flats, storage space is limited. We found the tops of wardrobes, which are usually too high and out of the way for everyday use, are ideal. Dried food is so light in comparison with other stored foods that it is a simple matter for the average housewife to handle and lift large containers, even when they are stored overhead.

Many housewives who move into smaller houses or flats when their children leave home, give up making jams and jellies because they don't have storage facilities. For them, dried fruits are the perfect solution. As we explain in Chapter 14, p 126, you can easily make up a quart or two of dried fruit at a time, then either store it in refrigerator or use regular bottling lids and keep it in a cupboard.

This not only eliminates the strain on shelf and storage space, but also means you do not have to stand for hours over a hot stove during the summer when fresh fruit is in season. Jams, jellies and preserves made from dried fruit are as good or better than those prepared conventionally.

The only foods you might not want to dry in the confined space of a flat because of the smell, are fish, poultry and onions.

But most other foods will fill your home with a marvellous aroma, better than any commercial air freshener. Although a back porch location for the dryer would solve this problem of smell, I have kept a dryer in the corner of a bedroom, and in the kitchen.

Because you have so many choices of the size and shape of dryers, even a small flat is no hindrance to learning to use this way of feeding yourself and your family. One of the dryers I designed especially for the British home fits snugly into a corner of my hall and looks like a nice piece of furniture. Just make certain the dryer is placed where the escaping moisture will not cause inconvenience although I have never found this a problem.

Living in the city, away from where food is grown, naturally means higher food prices. If you are lucky enough to find space for a garden, or can get an allotment, you will find your dryer a real moneysaver. Another solution is bulk buying. It should not be too difficult to find another like-minded family to share the buying of crates of produce, thus saving both of you money.

An even better way to buy produce cheaply is to make contact direct with the growers. One woman we know makes regular trips every summer to fruit orchards within a 50-mile radius of her home. She brings back enough food for herself and several of her neighbours. They pay her car expenses and everyone benefits.

It is a good idea, however, to make arrangements with the growers in the early spring, rather than waiting until everyone is out looking for bargains in fresh fruit and vegetables.

There must be many older people with garden space who are no longer able to do as much with it as they used to and who would welcome trading the use of their land for a share in the produce. Why not contact senior citizen organizations, or advertise in the newspaper in early spring, offering to do some, or all, of the gardening and divide the harvest? Both parties would benefit.

As with any new experience, the initial introduction to food drying is of vital importance. An imperfect drying technique may result in below-standard dried foods, often tasteless and even mouldy. You might well become discouraged unless you first experiment on a small scale. We recommend that you read this book first, then ask yourself the following questions: Am I interested in a long-term (at least three years) programme of food preservation? Do I have access to farm or garden produce, or can I buy fresh food in quantity at prices below those in the local shops? Do the advantages of food drying make it worthwhile for me to learn how to use an alternative to freezing and bottling?

If you can answer 'Yes' to these questions, then the next step is to experiment with small quantities of food, with a simple dryer. We believe you will eventually want either the box dryer, or the one especially designed for the UK, as described in Chapter 5. The plans for both of these dryers are simple enough to be followed by almost anyone handy with tools. You can, of course, ask your local carpenter to make one up for you. The chances are, judging from my own experience, that he will make one for himself as well when he realizes the advantages.

We know a young man who made a box dryer for himself, and ended up making them on a regular basis for neighbours, friends, and friends of friends. The clothes dryer model can be ordered from the address found in the Appendix, and while it is not quite as satisfactory as the box dryer, we found it did a very adequate job. It is extremely lightweight and portable, and dries clothes quite well too!

Before investing time and money in a larger dryer you may want to try one or more of the simpler styles mentioned in Chapter 5. Single trays with folding legs are easy to make and work well for very small batches. They are especially useful for drying a trayful of herbs at a time. For anyone with limited space, who keeps his heating system going most of the time during the chilly hours, this is a convenient and inexpensive method.

One of my neighbours, during the walnut season, places small trays like this over every heating area in her flat. She buys surplus nuts from a nearby grower for a very small cost, and keeps her heat going until the nuts are dry. Since nuts are harvested when the cool autumn days are upon us, this thrifty woman keeps her house warm and dries plenty of nuts for the winter at the same time.

Sun drying is, of course, the traditional form of food drying, but it is not very practical for the city dweller. Pollution tends to block the sun's rays, and may also affect the flavour of the food.

However, if you do have some warm, sunny days and have

food to dry, you might consider setting up a solar dryer on your balcony, in your garden, or even on an accessible rooftop. But remember the hazards of automobile fumes and also watch out for stray dogs and cats.

Unless you live on the ground floor, setting out drying trays in the garden will involve a lot of trips up and down stairs to stir the contents of the trays. If your garden cannot be protected from the smoke of automobile exhausts, however, you would be better advised to forget about sun drying. Build a box dryer and then you can be sure your food will not be contaminated by the waste products of industrialization.

Do not overlook the advantages of converting an old refrigerator or freezer. They can be easily modified to make very satisfactory dryers, can be installed on a porch or in your kitchen without looking out of place, and can be used for food storage when not used for drying. Their main disadvantage is that if you have to move them, they are very heavy. However, they are an easy way out of building the shell of the dryer, and sometimes you can pick up an out-of-date model from an appliance dealer or a friend for little or no money. (See Chapter 5.)

We have already pointed out the disadvantages of such methods as sun drying and drying in small quantities. This does not mean that many people don't use these methods, and usually very successfully. But being basically lazy (I prefer to call it 'efficient') and because I firmly believe that efficiency adds to the pleasures of living, I recommend the box dryers.

You can of course alter the specifications of the box dryer to fit the space available in your flat. Also, you might consider putting your dryer on rollers to make it easier to move about. And why not paint the dryer to match your colour scheme, or decorate it with contact paper, or even wallpaper it! I use the top of my dryer as a bookcase during the winter and spring.

The advantages of home food drying for the flat dweller, can be summed up as follows: less bulk, easy to transport, easy to store, all real advantages for families who may have to move; a home dryer is cheap and easy to build and operate, not as bulky as a freezer, much less expensive, food keeps longer, and it is easier to move if necessary; storage is crucial for most flat dwellers, dried food takes up only one-fifth of the space of frozen or bottled foods; bulk buying during the harvest season means greater food savings (when I came to Great Britain from the US I brought a year's supply of dried fruit and vegetables in ten 5-gallon tins!); the dryer can be adapted in size to fit into almost any flat; several families could share a dryer—this would cut down on the initial cost and encourage buying in bulk, thus saving on food bills and getting maximum use from the dryer.

The converted clothes dryer in Chapter 5 is probably the best type for the flat dweller, especially if you know you will be moving within a year or two. It does an excellent job, folds for compact storage, and is lighter than the box dryers so is easier to move. But if you are an enthusiastic handyman, make the smaller box dryer and put it together with screws. Then all you need do is disassemble it, pack it flat, and reassemble it when you are in your new home.

I know people who purchase used school lockers, the metal type, and use them to keep their dried food safe from insects and rodents, yet handy for everyday use. If you live in a flat without much cupboard space, this may be an ideal solution.

But no matter which dryer you choose, or how you obtain the fresh food, you will find that dried food will make housekeeping in the limited space of a flat much easier.

9 Home Dried Baby Foods

Commercial baby foods contain certain additives and are expensive so some mothers prefer to prepare their child's food themselves. Home dried food can make a significant contribution to your baby's diet. Because you take only the exact amount you need from the container, nothing need be wasted. Because it is so easily reconstituted, you can simmer the baby's daily portions of fruit and vegetables while you prepare breakfast. Then a quick whirl in the blender, a dash of seasoning, if you really consider it necessary, and the baby's food is ready.

Naturally there may be times when tinned food is needed, especially for busy working mothers, for whom such food can be a real help. But with your own home dried foods you can easily prepare a wide variety of fruit and vegetables your child will relish, at a much lower cost.

Carrots, peas, green beans, squash, beets, greens, potatoes, all these are in plentiful supply during the summer months. They all dry well and are palatable and good for a baby. When you plan your drying programme, keep your baby in mind, and make sure you select foods known to be important for the young child. Although some of the above-mentioned foods are fibrous, liquidizing, and then straining for the very young baby, make the food easy for small digestive systems to handle. When your baby begins to get teeth, he can probably handle slightly more solid food, but of course you can get advice about this from your doctor or health visitor.

When you prepare the family meals—and you can usually count on using at least one or two dried foods once you have built up a good supply—simply take a little from each dish, liquidize and feed to the baby. Usually you can make enough to store in the refrigerator for another meal if you wish. I used baby food jars for this, which made it easy to determine how much food the baby was eating.

Because your home dried food is not subjected to the high temperatures used during the processing of tinned or bottled food, it retains more vitamins. It is for this reason I recommend that you always simmer, and never boil, when reconstituting or cooking dried fruit and vegetables, although I do recommend pouring boiling water over the food for soaking as this in no way impairs the nutritive value and shortens cooking time.

When preparing your baby's food, the less water you use the better, as many vitamins and minerals are water soluble. Use left-over water to make soup with powdered vegetables and perhaps a bit of nutritional dried food yeast.

While this book is written primarily to explain the advantages of home dried foods, it may not be out of place in a chapter on baby food to mention the advantages of this modern wonder food. This is not the same as baker's yeast, but is grown from special cultures (originally from beer) and is exceptionally rich in vitamins, minerals and proteins.

Many people are surprised to learn we can get all our B vitamins and essential proteins from food yeast. The following chart compares a highly nutritious food, beef liver, with this yeast.

Brewers yeast (Powder)		**Beef liver**
100g		100g
Tryptophan	738mg	296mg
Leucine	3380mg	1819mg
Isoleucine	2520mg	1031mg
Lysine	3450mg	1475mg
Valine	2879mg	1239mg
Methionine	894mg	463mg
Threonine	2484mg	936mg
Phenylalanine	2010mg	993mg

So if you are thinking about a balanced diet for your baby, consider the addition of *very* small quantities of nutritional yeast to his cereal and vegetables, soups and stews.

Ease of preparation of dried fruits is a feature busy mothers will appreciate. In many cases, such as fruit, you do not have to cook it at all. Either soak as directed and when soft, blend for a quick, delicious and natural dessert. Or when the baby is old enough let him chew on a piece of dried fruit, instead of processed, devitalized food. You will be helping to instill eating habits which may keep him in adulthood, from eating synthetic foods to the detriment of his health.

In the chapter on Recipes, you will find instructions on making fruit leather, a superb baby food. Also, try sesame crisps and oatmeal crunch as a substitute for teething biscuits. They cost less, contain healthful ingredients, are easy to make, and are foods the whole family will enjoy.

10 Dried Food and the Freezer

I am often asked how dried food compares with frozen. And can the two techniques be used at the same time for maximum advantages from both methods?

Drying will probably never supersede freezing, unless soaring electricity costs make operating a freezer prohibitively expensive. Many people will continue to prefer the convenience of frozen foods, and feel that the added expense of purchasing and operating a freezer is worth it. I personally believe that the two techniques are extremely compatible.

Meat takes up much of the space in the average freezer. Home-drying of meat is not recommended by some home economists because of the danger of possible food poisoning if drying is not carried out properly. There is certainly far more danger than with fruit and vegetables, because meat is more susceptible to spoilage by harmful bacteria.

A resourceful drying enthusiast wrote to me saying:

> We used to smoke our meat, but now we find that if we use smoke-flavoured curing salt we can dry meat in the dryer with very good results. We then put the meat in our freezer, because it might still spoil without refrigeration. But it takes up much less space, and it is really good to eat.

I have friends who are hiking enthusiasts. They dry cooked pork chops, then freeze them, to take along on weekend camping trips. This is all right as long as the meat is used within two days of being taken out of the freezer, the weather is not too warm, and the meat is thoroughly cooked before eating.

During the summer months you may get an unexpected opportunity to buy a quantity of food you want to freeze, but the freezer is full. You can often remove surplus fruit or vegetables from the freezer and, after defrosting, dry them. I must emphasize you will not get top quality food as a result but let me quote a woman who raised eight children and learned to use every scrap of food around the house.

'I know this isn't as good as drying fresh food, but it's a lot better than going hungry,' she wrote. 'I've tried drying almost every kind of fruit and vegetable I can get my hands on. Some I haven't served yet, but I figure they don't take up much space and, who knows, maybe someone will really need them badly some day. They will keep just about forever if I make sure they are airtight and in a cool dark place.'

If for some reason you have to take food out of the dryer before it is ready to store, you can quickly pop it into freezer cartons and put in the freezer, When you have more time, just continue with the drying as usual. If packaged, nearly-dried food has collected moisture from the freezer, it just needs a bit more drying time.

But for real ingenuity, I think this thrifty woman must take top honours:

> I am one who will not throw anything away. At least, not if I can figure out how to make it usable. A few years ago we made arrangements to buy chickens from the poultry man when they were clearing out their older hens. These stewing hens were taken to the processor for dressing for our freezer. Sometimes we were able to sell some to friends when we had a surplus.
>
> The processor saved the eggs he found in the hens, most of them soft-shelled, which I then proceeded to use in the following way: I took them out of their shells and froze them, but I accumulated so many they were taking up room I needed for my summer blackberries.
>
> So I took them out of the freezer, defrosted them, then beat them up just a bit, and spread them on flat baking tins. They were then put in the dryer, and when completely dry (they get a bit crumbly) they went into plastic bags and back into the freezer. They took up much less space, and I use them when I want to bake. I sometimes add a bit of this dried egg mixture

to our dogs' food, so even our pets benefit from our dryer-freezer combination.

I have already mentioned (page 89) the convenience of cooking up just a small batch of jam from your dried fruit, then storing in the refrigerator or freezer. This cuts out the bother of using sealed lids or other airtight tops. It certainly is an improvement over having to handle hundreds of jars as I used to every year during the fruit season.

If you have a big family, the freezer-dryer combination will keep them supplied with your good home-preserves and at the same time make the work much easier.

Many people wonder what the difference in convenience is between dried and frozen foods. The answer to this depends on your individual circumstances and requirements. Reconstituting time for dried food is about the same, or perhaps a little less, as the time it takes to thaw frozen packets of fruit or vegetables

Simmering dried food will reconstitute it in from 15 minutes to an hour, depending on the food. It may take up to 24 hours for reconstitution by soaking in water. Remember, however, much depends on how dry the food is. The drier the food, naturally the more water you must restore. But if you drop the food into boiling water without first reconstituting, then the cooking time should be about the same as for frozen foods.

Freezers are superior, in my opinion for such foods as meats, or for frozen fruits when you want to use them for sherbets, ices or ice cream dishes.

What do I like especially about my store of dried foods? I don't have to worry about whether or not the freezer is working, or whether the power might be cut off. I don't have to take frozen food out to let it thaw, only to forget about it, as I have sometimes done, and find I have to throw it all away.

For travelling, camping, or for the boat-owner dried foods save time, money, space and weight. You can of course take some frozen food along on a one-day outing, if you are careful to put it in an insulated cooling box. But frozen foods must be used very soon after they have been defrosted, which limits their convenience away from home.

So both dried and frozen food has its own special advantages. Do not get rid of your freezer, thinking your home food dryer will take its place. Try drying food first. Be experimental and creative, use your imagination to make the two systems complement each other.

When you look at your rows of storage containers, filled with dried food, I guarantee you will feel a nice warm glow, knowing that you have food which will keep your family in an emergency. And best of all is the knowledge that you did it yourself. In these days when man sometimes feels he is only a little cog in a big, impersonal machine, it is gratifying to be able to take some positive action to improve one's life, with minimal dependence on the industrial establishment.

11 Travelling and Camping

Modern man measures his status in life by the amount of comfort and luxury he has achieved. We cherish our pushbutton economy with its genie-like ability to produce what we want when we want it.

Yet every year countless thousands of families leave their comfortable homes with central heating, soft furniture and all the convenience of gas and electrical appliances, to sit around a campfire cooking a pot of stew or a can of beans. Then they crawl into their tents or sleeping bags to spend a night under the stars.

There must be some powerful atavistic instinct that lures man out into the fresh air and wide open spaces. Whatever the reason, if you like a bit of luxury when you are 'roughing it', a camping holiday can be more fun if you add some of your own home-dried foods to your list of supplies.

The advantages of dried food are many for the camper or traveller. It is lightweight, making it ideal for backpacking; condensed in volume it takes up only a small fraction of the space of fresh or tinned food. It is virtually spoilage-free if kept sealed; and if home-dried, is free from chemical preservatives and additives.

We know one man, now well into his seventies, who still goes riding into the wilds every year. He carries his own home dried products as his main food supply. Some foods are dried during the winter, using the wood stove in his front room, which is kept going all the time anyway during the bitterly cold winter months.

He says he enjoys his trips three times! First, the anticipation while the food dries, as he thinks of what he will be doing when he eats it; second, the actual participation in the trip; and third, the recollection of his experiences when it is all over and he is back home again.

In winter he dries such foods as broccoli, cabbage, carrots and also nuts and late apples. He hangs a screen from the ceiling near the door jamb, and always keeps a window open, about 5cm (2″) at the top, to make sure there is a moving current of air.

'If your food moulds,' he suggests, 'you may not have the window open all the time, or the room is [perhaps] too cold. But I like winter drying because the heat costs next to nothing, and it keeps me thinking about what I'll be doing next summer.'

You can prepare meals in individual packets for each day you plan to be away. Soups, stews, casseroles, all can be made up from your stock of dried food. Each night open a packet, add whatever else you may have on hand, and you have the makings of a healthful, easy-to-fix meal, quickly prepared.

Some Camping Menu Ideas

Powdered vegetables for instant soup. Experiment to find the combination your family enjoys best. My family like a mixture of powdered onion, potatoes, carrots and parsley. Incidentally, if you can't powder the vegetables because they aren't quite that dry, they'll turn into flakes and the soup can then either be strained or eaten as a thicker and heartier 'instant' soup.

For added nutrition, add a bit of nutritional powdered yeast (always ask for the debittered kind, available in health food stores, it is much better tasting than the bitter yeast) to the dried vegetables, then pour boiling water into your cup for a quick, delicious and healthful drink. It makes a good substitute for the inevitable cup of tea or coffee.

Dried fruit makes a superb snack, especially good to carry on a hike through the woods. And it is much better for you than sweets, because it supplies naturally good flavour and quick energy.

Keep a supply of dried fruit with you in the car during long trips for a quick pickup. Many people suffer a sudden drop in blood sugar when under stress or overtired. This can lead to a lapse in concentration with sometimes unfortunate results. Keep dried fruit in the car to munch on when you feel a bit tired. It might help avert a serious accident.

For a quick and easy dessert, soak dried fruit in the morning for a dish of stewed fruit at dinner time. It is especially good with

Fresh peaches and dried peaches

Fresh French beans and dried French beans

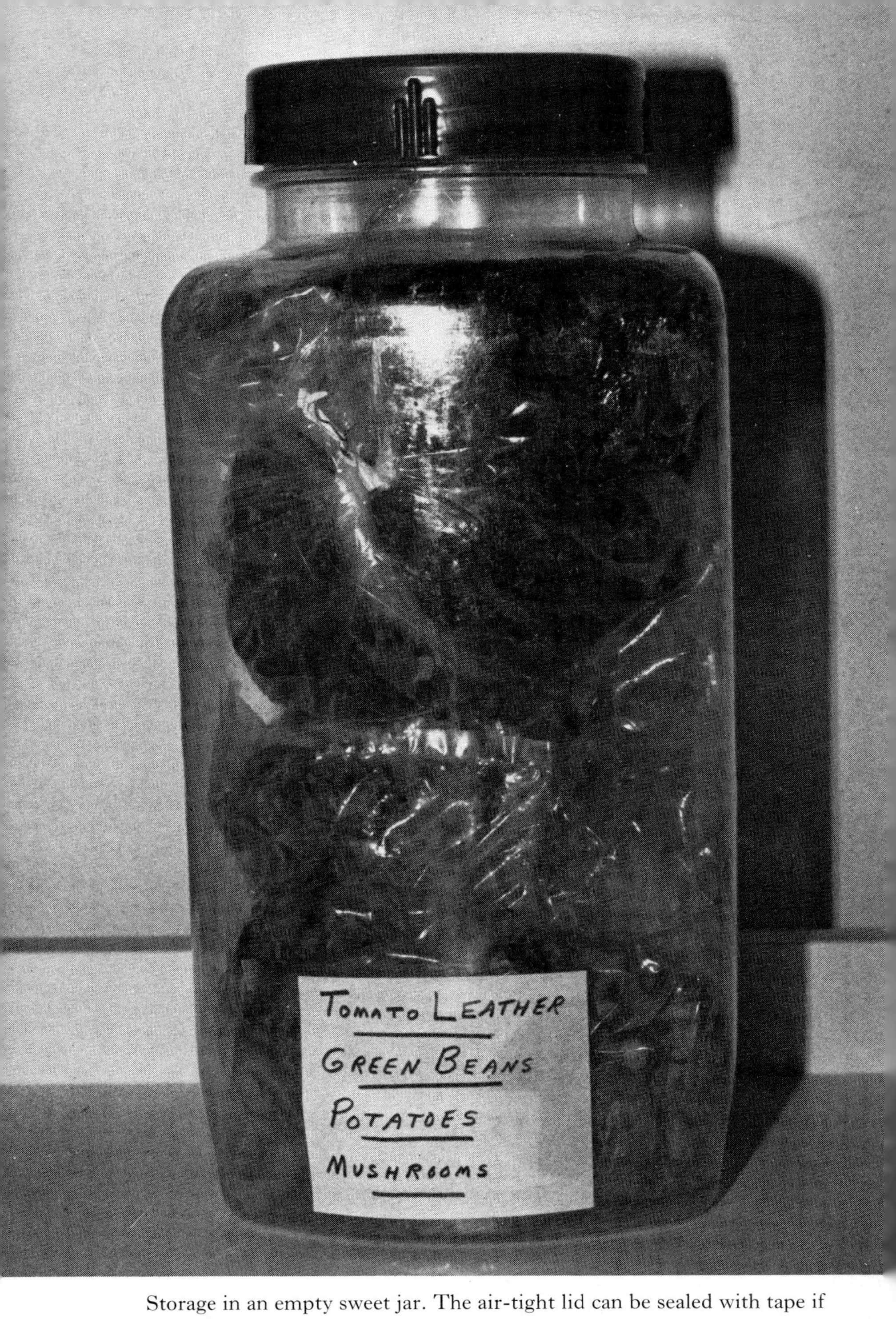

Storage in an empty sweet jar. The air-tight lid can be sealed with tape if desired

slices of fresh orange—leave the peel on to sharpen the sweetness of the fruit.

Surprise your family with a Chinese supper dish around the campfire one evening. In the morning, soak dried vegetables such as cabbage, carrots, greens and onions, and toss in a bit of fresh ginger root if you have some.

To prepare, drain off the soaking water, and heat oil in a cast iron skillet or frying pan. Add minced ginger (about ½ tsp or to taste), then stir in the drained vegetables and salt or soy sauce to taste When ready—vegetables won't take more than 5–10 minutes to cook, and should not be mushy soft—combine ½ cup cold water with ½ tsp sugar and 2 tbsp corn starch, or flour, to make a smooth paste. Add to the vegetables, and stir until thickened. Use the soaking water from the vegetables to make this gravy. Add any ready cooked meat or fish bits you have on hand. Serve with rice if possible, but this dish makes a good meal by itself.

We always carry with us ready-mixed rolled oats and chopped dried fruit for a quick breakfast. Pour boiling water over a bowl of this mixture, let it sit until the oats are softened. Add milk, sugar or honey for a delicious and quick breakfast. Sliced bananas are good on this, as well as any fresh berries you have been lucky enough to find.

If you have dried bananas, soak a few pieces right in with the oats. They may still be a little crunchy when you eat them, but will add real fresh-banana taste to the meal.

The Chinese considered dried snake a delicacy for travelling. It is regarded with equal enthusiasm by some residents of America's southwest, where it is sold in food speciality shops at high prices. I am told that it is rather sweet but must be chewed for hours before any appreciable amount can be swallowed.

In Tibet they make a unique convenience food from yak butter and dried powdered tea. These ingredients are blended into an oleaginous ointment, then stuffed inside hollow bamboo tubes. The ends are then sealed, and the traveller has only to unblock one end of the tube and shake out some of the mixture for instant nourishment.

Our family prefers a somewhat different mixture. We blend soya flour, salt and cold-pressed oil (preferably sunflower or sesame) until creamy. Add toasted sesame seeds and mix thoroughly. Eat as it is, or make it into a candy bar by adding melted honey and powdered milk, mixing well. This should then be poured onto baking sheets and when set, cut into squares and wrapped in plastic film for quick snacks.

Dried, finely chopped fruit is another good addition to this versatile combination. This not only tastes good but is high in protein and contains lecithin, which helps to keep down cholesterol.

Most convenience foods in the past were developed to help travellers stay alive while passing across bleak or dangerous territory. But *charqui*, a dried meat believed to have been developed first in South America, was used for keeping tribes in food during the lean winter months. The word is thought to be the forerunner of the American pioneer's 'jerky', a very hard type of dried meat still in use in the US today.

The South American natives made *charqui* by cutting beef into very thin slices, then dipping the slices in a strong saltwater solution, or simply rubbing the salt directly into the meat, which was then rolled up inside the hide of the animal whose flesh was being treated. It was left like this for twelve hours or so, when the salt would have been absorbed into the meat.

At the end of that time the meat was taken out of the hide and hung up in the sun to dry until it was as hard as the famous salted cod made on America's north Atlantic coast. *Charqui* was not easy to eat, as it was so tough, but it would keep a man alive, so it probably tasted pretty good to the hungry traveller. It could be cooked, but the natives had first to pound it between two rocks to prepare it for the pot.

Today, jerky sticks are sold throughout the US during the hunting season. They are probably much more palatable than *charqui* but both are essentially the same thing—a hard, dried meat which lasts for long periods of time and can be easily carried for emergency rations.

Modern campers might want to experiment with drying their own cakes and biscuits as some North American Indians did, although I doubt if the powdered flour they made from dried mesquite beans would be much to twentieth century tastes. The mesquite pods were pounded in a mortar into a rough flour, then mixed with water, placed in containers, usually pottery or woven baskets, and left in the sun to dry. These dried cakes were stored for food during the lean winter months. They were also carried by travellers as they kept well, weighed very little and, for variety, could be soaked and made into a kind of porridge.

You might want to try drying some of the cakes, biscuits and crisps to be found in the chapter on recipes. Any of the crisps make wonderful titbits for travelling. If your dryer is full, spread the mixture on baking sheets and dry in a very slow oven until quite brittle. Then break into pieces and store in paper bags. Carry on hikes, while mountain climbing, or keep by you to munch while driving.

See the section on drying meat, for ideas on how to use your dried meat and fish for camping and travelling. Just be sure to observe the precautions on drying and storage, as it is possible to get food poisoning if these products are not treated properly.

For some practical and innovative ideas while camping or on the trail, I quote again from our survival expert, Clint Seibert.

> I reconstitute my dried food when I first get up. Because I like raw, uncooked foods whenever possible, I usually put dried apples, raisins, then honey, wheat germ and a pinch of coconut, all into a container, fill with water and let it stand until I am repacked and ready to leave camp. Then I empty out the water and place the food in a container that will fit in one of my inside pockets. You can use a plastic bag for this, as the food won't be in contact with the plastic for more than a few hours.
>
> Tie it tightly so it won't spill. In another pocket I put some dried bread and raw Spanish jumbo peanuts. Also, always keep a tablespoon in a pocket at all times.
>
> I find as I ride along, the heat of my body, plus the jostling as I walk or ride my horse aids in reconstituting the food.

Large thermos flasks are excellent for soaking food. They shouldn't leak, and are usually sturdy enough to stand up to a rough trip. You can prepare your food in them in the morning and, when you are ready to make camp that evening, your food is ready to be cooked, if necessary, or eaten as it is.

Clint Seibert's Dried Eggs

Break a dozen eggs into a bowl, stir, pour into a glass baking dish, and season as desired. Bake until a straw inserted in the centre comes out dry. Be sure the eggs are thoroughly cooked. If they are not, the partial cooking can cause food poisoning.

Cool, then chop in 6mm (¼″) pieces, and place in dryer. When finished, pieces should be so dry they rattle. Place in a brown paper bag, leave the top open, and hang in dry room for a week to season. Seal in a metal container. Eggs dried like this will keep for years. Change the paper bags every six months or so, because the oil will slowly soak into the paper. As you change the paper, this helps to get rid of the oil which is one of the things which may lead to food spoilage.

Clint Seibert's Dried Meat Loaf

Grind *lean* meat, season to taste, form into meat loaf 5cm (2″) high and leave 12mm (½″) gutter around edge of loaf. Drain the oil that gathers there as the meat cooks. Roast in a moderate oven until thoroughly cooked. Cool, chop into 6mm (¼″) pieces, dry. Season in open-topped brown paper sack for about one week. Change oil-soaked sacks every six months or so, as with eggs. Meat dried like this will keep for years.

12 Dried Food in Emergencies

One of the hardest jobs in the world must be to sell insurance. No one wants to think about unpleasant possibilities like illness, death or such catastrophes as war, famine and other disasters.

In the early 1970s a dam burst in a section of the northwest United States. A friend who had relatives living there described to me what happened to those people who escaped with nothing but their lives and the clothing they wore. These people belonged to a religious sect which advocates keeping a year's supply of essential food and necessities on hand, for every family in the congregation.

When the dam burst food and other supplies were instantly available, to help those who had lost everything. These people didn't have to wait for relief organizations to move in and help them. There was a planned, instantly operative programme to supply relief and emergency supplies when needed.

Our family lived for a few years in a remote mountain area with unpaved roads, and where stormy weather was a matter of course. The mail was delivered only when the postman could get through with his four-wheel drive vehicle. We had to keep a good emergency supply of food and fuel on hand at all times.

We had a large freezer, but the power lines were often down, so we learned to depend on food which could be stored without electricity. The years we spent in this American wilderness had their enjoyable side, of course, as any life close to nature must. But it also taught us the necessity of keeping on hand non-perishable food.

Modern man is so accustomed to walking into a shop and picking whatever he fancies off the shelf that we find it hard to imagine a time when this would not be possible. But we live in a fragile and vulnerable environment (witness the power failure that blacked out New York City in the late '60s and the miner's strike which caused black-outs in London in the early '70s). Such failures can disrupt an entire system of supply for a community, and at these times a good supply of dried food in the larder would be a great comfort.

Ask yourself these questions: what would I do if no food were available in my local shop, no petrol for the car, and no electricity to keep the freezer operating? How long could my family survive under these conditions?

Then imagine your cupboard with a year's supply of dried vegetables, fruits and perhaps a few other staples such as salt and spices. You would feel much less panic, and the crisis, whatever the cause, could be faced with more equanimity. *Passport to Survival* by Esther Dickey (Bookcraft Publishers, 1848 2300 South, Salt Lake City, Utah, USA) gives a remarkably complete and feasible plan for families who wish to prepare themselves for an emergency situation.

Summer is the time to start thinking along these lines. Produce is more readily available, with better prices, better quality and greater variety. I know many families who dry all they can get. I usually end up with lots of apples, as we live in apple country, but also because my family enjoys apples in any form.

As for vegetables, for this emergency storage programme you should consider drying potatoes, even though they store well in a cellar. But we are talking now about preparing for unusual situations. Dried food is lightweight, which would make it much easier to transport if you had to leave your home in a hurry. If it is stored, as it must be, in airtight, watertight containers it is much less likely your potato supply would be destroyed, as in the case of earthquakes or floods.

In the meantime, if the emergency never happens, and of course this is what we all hope, you are supplied with food for the coming year. You can't lose with a plan like this!

Sit down and work out what foods are usually available in your area during the summer harvest. Then budget so you will have money to buy extra food, always remembering you are paying less now for the food than you would during the winter.

Begin collecting all the storage containers you can find. Ask your friends to save their plastic ice cream cartons for you.

Square shapes are better than round because they waste less space. Three-pound coffee tins are good. Any other large containers that will keep out air, moisture, light and insects are what you are looking for.

Always keep a list of the dried food you have stored, and always label each container carefully with the date as well as contents, just as you would for freezing or bottling. I mark my containers twice, one side with a felt tip pen, the other with a plastic label, in case one or the other is removed by accident.

Make up soup mix packets and put in extra potato. The starch will thicken the soup and give it a little extra body that may be appreciated if rations are short. I also add substantial amounts of onions and never omit herbs. Bay leaf, thyme, marjoram or sweet basil are good. Herbs will brighten up a monotonous diet amazingly.

Dried fruits by themselves will provide energy and help to satisfy a sweet tooth. Pack one 5-gallon tin with a variety of foods. In a crisis a younger child could carry enough food for the whole family for several days in one such container. This could prove unexpectedly useful in case of a sudden need to leave your home.

These ideas may seem far-fetched and scarcely worth your consideration. Those of us who have always lived in a safe, secure society find it hard to imagine living under conditions of extreme hardship and deprivation. But all over the world, especially in areas hit by natural or military disasters, these conditions are commonplace, and the suffering can be intense.

You really have everything to gain and very little to lose by preparing for an emergency. If it never happens, your stored food will always be useful. You may even be able to help some friend or neighbour should they suffer some unexpected disaster.

13 Food Drying Round the World

Even in lush tropical countries where one would imagine there is always a steady supply of fresh foods, the native peoples dry food for storage. In Sri Lanka (formerly Ceylon) fruits, vegetables, nuts, meat and fish have been dried for centuries.

Yams, for example, are first boiled, then slit and dried in the sun for three days. Then the fibres are pulled out and the food is dried again. It is spread out on pieces of rush matting and then covered with nets to keep the birds away.

Another variety of yam is dried without boiling, then pounded into flour for use whenever starch is required. It is especially good for steamed dishes, such as dumplings to put in soups.

Fish can be dried either with or without salt. The less oily species are preferred there such as thora, mullet, and the small sharks, but not the man-eating species, my Sri Lanka friends assure me.

Most fishermen sell as much of their catch as they can, then dry what they have left for their own use later in the year when monsoons prevent fishing. Rush matting is spread out on rocks, the fish laid on top, and then the children keep watch to scare the birds off.

In the jungles of Sri Lanka the natives dry meat in much the same way as the American Indian dried his pemmican and jerky and the African bushman his biltong.

Chillies, both hot and sweet, are a popular food and are often dried. First, they are boiled to break down the tough outer skin, then they are spread on matting to dry in the sun. In addition to yams and chillies, the Sri Lankans dry a delicious candy-like concoction from the fruit of the palmyra tree.

The outer fibrous cover is removed from very ripe fruit, the inner part is extracted and spread out on mats to dry in the sun. Each day another thin layer of the fruit is spread on top of the previous day's layer, until after five days it is half an inch thick. It is then left to dry completely, when it turns a creamy light tan in colour. It is removed from the mats, cut into slabs and rolled up, something like the fruit leather we make. (See page 77.) Then it is once more put out in the sun to make sure it is thoroughly dry. This makes a wonderful sweet, and will keep a long time if stored properly.

Sometimes the dried fruit is dipped in a sweet syrup and dried yet again, then stored. Our Sri Lanka friends say it tastes amazingly like chocolate.

Dried food is generally stored in boxes woven from the leaves of the versatile palmyra tree. These boxes are square in shape, with tight-fitting covers, and can be stacked one on top of the other, then hung over the fire-place, where food can be kept for years in good condition.

Even the seed of the palmyra fruit is used. It is planted just to harvest the root, which looks a bit like a yam. When the seed has become as large as a big carrot, it is split, dried in the sun, and kept as a vegetable.

We asked about weevil infestation, a problem likely to attack all dried food. In Sri Lanka they inspect the food about every three months. If there are any signs of weevils, such as cobwebs or larvae, the food is put out on the rush mats to dry again. We recommend the same regular inspection of your own dried food in much the same way you might make regular inventories of the contents of your freezer. If you can catch weevils at the very beginning of their cycle it is easy to oven-sterilize the food.

It was while doing this with a package of dried apples that I discovered a new way to serve them. I had opened a packet of apples to make apple sauce and saw a trace of cobweb, so I discarded the pieces of fruit which seemed infested and popped the rest into the oven on a baking sheet. After 15 minutes I took the apples out, inspected them and saw no signs of worms, so I tasted an apple and found it was really delicious just warm from the oven.

Then later in the evening, I took a bowl to munch while watching television, and found that as the apples had cooled they had become crisp, and were delicious—but in a slightly different way. That's one of the exciting things about dried food, you keep discovering new ways to prepare and use it.

In Bavaria is found a small variety of pear, called *most birren.* These small russet pears are famous not only for the perry (pear cider) made from them but also as an ingredient of a luscious Christmas bread. (See Recipes: Birrenbrot.)

A young agricultural student while studying in England told us how he used to steal the pears from the oven as they dried. 'We would keep taking one at a time, until finally our parents would notice the fruit was disappearing, and then we would be forbidden to enter the kitchen. But they were so good as they began to dry, it was impossible to resist.' These pears are used to make 'birrenbrot'. The pears are believed to provide a superior food for the yeast used in the bread. After being dried, they are stored in ordinary cloth sacks. Because of the long cold winters there is no fear of weevils.

This same student spent a few years working in Madagascar, where he saw coffee spread out on the streets and pavements to dry. Chickens were turned out into the coffee beans to scratch among them, eating all the stray cereal seeds that might have made their way into the coffee and, at the same time, turning the coffee over which helped it to dry more evenly. I'm afraid, however, this is a method we westerners would find less than acceptable!

While in India he saw Bombay duck being dried out on the beaches. This is not duck at all but a fish called bummalo, similar to salmon, which is dried in Bombay and exported all over the world.

In pre-war Poland dried peas were an essential part of the diet of the country people. The peas were left to mature, then spread on wooden trays to finish drying in the sun. (We recommend the same procedure, but if you have a dryer it is faster with less trouble.) The peas were always dried pebble hard, and were soaked for three days before cooking. Broad beans and runner beans were treated the same way. Naturally the fresh vegetables were eaten during the summer, but people always planted a surplus in order to have enough to dry for the winter.

Soya beans were also an important food for pre-war Polish

villagers, and they, too, were dried and stored. Woodland varieties of mushrooms were popular, and were gathered by the basketful and dried. If you make a careful study of wild mushrooms they can add a delightful variety to your winter menus. Please always make certain you have some reliable reference books, with plenty of colour illustrations, to help you make absolutely positive identification of every mushroom before eating. Even then it is wise to eat only a small amount when first tasting a new variety.

Fish in Poland were first salted, then fastened to poles or ropes and left in the sun to dry. They were then hung in the attic for winter storage. In the old days plums, pears and apples were dried either over the oven, or in the oven after the day's bread baking was finished.

A smoke house could be found on most Polish farms where the famous Polish sausages were made. Plums and pears would be smoke-dried in the same smoke house. After being boiled the dried fruits were generally used in special Christmas cakes and puddings. All kinds of game, especially venison, were boiled, salted and smoked, then hung in the attic. Only pine wood was used for smoking by the farmers, who preserved food in huge quantities, selling much of it in the markets.

My Polish friends say their mothers never used vinegar to preserve foods such as mushrooms, instead using citron. It was considered to produce a superior product, and I pass the idea along to any adventurous cooks who may care to try it.

In Iceland, two foods which dry well are seaweed and fish. The special seaweed, known as 'dol' (dulse), grows on rocks near the mouths of rivers and in lagoons. Although it may seem a strange food to western tastes, in Iceland and throughout the Orient dried seaweed is considered a delicacy, and is known to be very rich in minerals and vitamins.

It is first picked off the rocks, cleaned in the sea, to remove the green algae and other plants that grow on it, then spread on roof-tops to dry. The roofs are made of corrugate metal, a perfect material and shape for wind drying. When it is dry on one side it is turned over. When finished, it is placed over a barrel and pressed flat to force the salt to the surface.

This seaweed is eaten either plain or with butter, but always raw. One way to eat it is to scrape the salt off with your tongue, then eat the rest. In the Orient wonderful soups and vegetable dishes are made with seaweed. I often buy it in speciality markets and serve it to our family.

An old Icelandic legend tells of a poet named Egill Skallagrinsson whose beloved son was drowned. So overcome with grief was the father that he shut himself up in his room and determined to starve himself to death. Finally, after several days, his worried daughter tempted him to eat just a little 'dol', having convinced him it was not really a food. The poet agreed to do so, but the seaweed made him so thirsty he had to ask for water, and this was the end of his determination to commit suicide.

Eating 'dol' is a very old custom, and many Icelanders are happy the practice of gathering and drying it for domestic use is coming back into fashion again.

Cod, haddock, catfish and halibut are dried in Iceland although catfish and halibut are sometimes smoked because of their high fat content. Generally the fish are filleted only when dried commercially. The fish is split, then hung in a special drying shed called a 'hjallur'. This has no wall around the lower half, to permit the wind to blow through, but has a roof to keep out the rain. The shed can also be used to dry clothes. The fish are hung up on poles; the head is cut off and dried separately. The eyes are always eaten, as it is well known that this part of the fish contains valuable nutrients.

The fish take about five months to dry, from January to the end of May. Dried fish is quite expensive to buy so the people on the west coast of Iceland usually dry and eat their own. It is so hard that often it must be hammered into pieces to make it easier to cook. The slab-like pieces of fish are stored in great stacks in sheds. This type of food has long been a valuable source of protein for Icelandic people during their long, cold winters.

But in every country we have visited, every person we have asked agrees that food drying is practised and is considered to be the oldest known method of food preservation still in use. Every nation has its own special ways, its special methods and secrets developed over the centuries to make drying food a successful and practical way to feed its people.

But interesting as these exotic practices may be, what relevance have they to you and me? At first glance apparently none. In our highly specialized and technically advanced society, self-reliance and self-sufficiency are not widely practised. Until the present time drying our own food could only be regarded as a regression. But the world population explosion, combined with the rise in oil prices, has produced food shortages that affect even the affluent countries. Scientists and economists warn us that it is unlikely low food prices will ever return.

As we have seen, in the past food was dried in order to make good use of all the available food supply. We can no longer continue with our present wasteful lifestyle, with no thought for future generations. It makes good sense, therefore, for every one to investigate the feasibility of an alternative like home food drying. This could not only help individual families to enjoy better and more healthy food, but also help the nation's economy by cutting down on the vast amounts of food imported every year.

By using modern improved dryers (see Chapters 4 and 5), we no longer have to contend with the dust, bugs, dirt, rain and birds as do those using more primitive methods. Modern home food drying uses almost no fossil fuels, in sharp contrast to the huge electricity demands of freezers. Tinned food needs enormous quantities of energy to run the machines and factories where the food is processed.

Home gardeners can now utilize every bit of food they grow. How often have I heard friends say, 'I grow more food than I can ever use. Every year I give it away to friends and neighbours. If only I had all this food during the winter when prices are so high.' Your home food dryer will solve this problem beautifully.

All you must do is make up your mind you want to try food drying. Start with a small dryer at first, and prove to yourself that it can help you lead a better life. With a minimum of equipment, time and money the way is open for almost anyone to improve living standards and bodily health through the techniques described in this book.

14 Recipes

Now that you have dried food, what do you do with it? As you experiment, possibilities will continually suggest themselves to you. The following basic rules will help you to get started.

Fruits

Dried fruits and dried fruit leather can be eaten just as they come out of the storage container. To reconstitute for use in such dishes as pies, cobblers, jams, sponges, etc, use *one* of the following methods:

1 Leave packet open overnight so moisture from the air can soften the food. This works best if you open the packet on a rainy day when the humidity is high. Sometimes just exposing the food will soften it sufficiently to make it moist and chewy enough for eating.

2 Soak food in boiling water, just enough to cover, until food is of desired softness.

3 Simmer in covered pot (drop into boiling water but immediately lower heat to simmer) until tender, usually between 20–30 minutes. Do not add sugar to the food until it is cooked, as sugar interferes slightly with the dried fruit's absorption of liquid. Never throw away the water; it not only tastes good, but contains valuable nutrients.

Add more water if necessary during cooking. We sometimes add a little more than we need because the cooking water is delicious when poured over cereal, either instead of or in addition to milk, or added to fruit punch. It is even a nice tasty drink all by itself. But don't use too much water or some of the flavour will go from your fruit.

The rule for reconstituting fruit: 1 cup dried fruit = 1½ cups reconstituted fruit.

Time required: 1 hour up to 24 hours, depending on which of

the above methods you use, how dry the fruit is, how large the pieces, and what kind of fruit you are using.

Since some vitamin C is lost through processing, we sometimes like to add a squeeze of fresh citrus fruit, either grapefruit, lemon, lime or orange, to restore some of this vitamin. It also gives a pleasant piquancy to the fruit. Add just before serving, not during cooking.

Use reconstituted fruits for almost any dish where either fresh or cooked fruit is called for. Dried fruit when not reconstituted is good when chopped or ground and added to cereal or for making granola. Berries can be powdered and used to flavour icings, milk shakes, jellies, fruit punches and topping.

Vegetables

Most vegetables require reconstituting in water before eating. Some exceptions are courgettes (zucchini), carrots, and cucumber chips, which are nice for party dips or eaten alone as a healthful, low-calorie snack. Dried herbs and celery leaves are of course widely used for seasoning. Maize (sweet corn) can be eaten when prepared as you would the fresh variety, or ground into corn meal for many nourishing and tasty dishes.

To reconstitute:

Pour boiling water over vegetables, then let them soak until near normal size. This shortens the eventual cooking time which of course helps to preserve nutrients. Three hours soaking time is usually adequate.

How to Use Reconstituted Vegetables As side dishes for meals. In soups, stews, casseroles, topping for pizza (mushrooms, peppers, onions, tomatoes especially), powdered and added to yogurt, cream cheese or cottage cheese for dip mixtures or salad dressings in omelettes and meat pies.

Soups and Stews Dried vegetables can be dropped right into the boiling water of the soup pan without reconstitution beforehand.

Carrots will reconstitute in cold water in 5–6 hours.

Maize (sweet corn) should be cooked in equal parts of water, and tastes just like the fresh vegetable.

Omelettes: Reconstitute vegetables first, drain thoroughly, and use as you would fresh ones.

Casseroles: As with omelettes, then bake in moderate oven. See **Basic Vegetable Casserole.**

Substitution of Dried Vegetables for Canned or Fresh Substitute one-half the amount of dried food for a portion called for in ordinary recipes. (This is only a guide, as the amount must vary according to how much water is left in the food after drying.) An exception is green pepper which requires only one-fourth the amount.

In winter use your out-of-season foods, such as tomatoes, peppers and green beans to supplement your meals. You can dry complete soup or stew mixes. I combine potatoes, carrots, onions, celery, tomatoes and parsley as a base. If I have turnips or parsnips, in they go. Package in proportions you prefer in your soup.

Dry the vegetables separately (on separate trays) and then combine them in meal-size packets. Next winter you will have ready-mixed soup on your storage shelf, a great beginning for quick, easy but nourishing suppers, with most of the preparation done in advance. Add what you have on hand in the way of fresh vegetables.

To thicken soups, dried powdered potatoes make a great addition (or use unpowdered, just adding a bit more than the amount you would usually use). Ground brown rice flour (brown has more vitamins, minerals and natural roughage than the white rice) makes a nice thickener. Another interesting alternative, one I use to thicken a Quiche Lorraine, is to grind sunflower seeds, then stir them into the egg mixture. You get extra flavour plus a big bonus of protein.

Another excellent use for your dried vegetables is to blend them into a powder for a vegetable broth base, the exact proportions depend on individual tastes. Some people can't abide any garlic, others won't cook without it. My family likes a heavy concentration of celery and onion, with potatoes and tomatoes. We have a friend who puts dried sweet corn into most of her soups and stews.

A tasty, low-calorie and highly nutritious cup of instant soup can be made as follows:

1 tbsp powdered soup mix.

1 tsp debittered nutritional yeast (like brewers yeast or torula yeast).

Pour boiling water over the mixture and stir until dissolved. This makes a very satisfying pick-me-up for the dieter.

Any food you powder must be absolutely brittle dry.

The following recipes should help you get started.

Empanadas

½kg (1lb) dried apples, peaches or apricots
900ml (4 cups) (32 fluid oz) water
2 tbsp cinnamon
450g (2 cups) sugar
⅛ tsp salt
1 orange or 1 tbsp frozen orange juice and 1 tsp dried orange rind
(*Note:* All cups are 2/50g (8oz)

Simmer fruit in saucepan until soft. Add cinnamon, sugar, salt and the grated rind as well as the juice of the orange. (You can use frozen orange juice and your own dried orange peel, according to taste.) Let mixture cool. Use any good biscuit recipe (a rich scone recipe will do) to make up 7/50g (3 cups), and put in an extra 125g (½ cup) shortening for a richer dough. Add 180ml (¾ cup) water, all at one time, stirring 25 strokes *only*. Knead dough gently, about 15 strokes, on a lightly floured board or cloth. Pat or roll out to 6mm (¼″) thickness. Cut in 10cm (4″) circles. Place 1 tsp of fruit mixture on one side of circle, fold over, pinch shut to seal. Glaze with a mixture of 1 beaten egg with 2 tbsp water. Bake at 204°C (400°F) for 20 minutes or until brown.

Instead of glazing, when removed from oven brush with butter, sugar and cinnamon. These little sweets may be frozen and reheated.

American Biscuit Recipe

675g (3 cups) flour
1½ tsp salt
4–5 tsp baking powder

225–280ml (1–1¼ cups) milk
6 tbsp butter, or other shortening, melted

Combine and sift first three ingredients. Then add milk, stirring briskly, until mixed. Finally, fold in melted shortening.

Dried Fruit Streusel Pie (with crumb crust)

500g (2 cups) any combination of dried fruit (cherries, raisins, prunes, peaches, apricots, apples)
90g (3oz) oil, margarine or butter
150g (5oz) flour
3 tbsp brown sugar
1 tsp cinnamon
1 egg
125g (4oz) biscuit or cake crumbs (US—cookies)
60g (2oz) chopped almonds
60g (2oz) additional brown sugar
1 lemon rind, grated (or 1 tsp dried)

Soak dried fruit in water to cover and simmer 15 minutes until fruit is soft but not mushy. Combine fat, flour, 30g (1oz) of brown sugar and cinnamon, and mix until firm. Refrigerate this mixture for 30 minutes. Mix crumbs, almonds, brown sugar and lemon rind. Spread fruit on bottom of pie plate, then cover with crumb mixture. Sprinkle top with additional brown sugar. Bake at 190°C (375°F) Mark 5, for 30 minutes or until fruit is bubbly and topping is golden brown.

Dried Apricot and Fig Pudding

250g (½lb) dried apricots
250g (½lb) dried figs
2 slices whole wheat bread
60g (2oz) brown sugar
2 tbsp lemon juice
1 tbsp additional brown sugar
1 cup rolled oats

Soak dried fruit to reconstitute until soft. Chop fruit or put through grinder. Crumble bread, mix with sugar, fruit and lemon juice. Place mixture in bottom of baking dish. Sprinkle top with mixture of rolled oats and additional brown sugar, dot top with butter, and bake at 190°C (375°F) Mark 5 for 40 minutes, or until top is lightly browned. If served cold, serve with whipped cream. If served hot, serve with plain cream.

Apricot Rolls

16 digestive biscuits (graham crackers) (See recipe, page 132)
125g (½ cup) chopped walnuts
250g (1 cup) dried apricots
2 tbsp butter or margarine
2 eggs
125g (½ cup) white sugar
125g (½ cup) brown sugar
1 tsp vanilla essence (extract)
250g (1 cup) grated coconut

Crush biscuits, add nuts. Chop apricots in blender. Melt butter in large heavy skillet. Add apricots and stir to prevent scorching until butter is melted. Remove from fire. Blend eggs and sugars in blender thoroughly, then add to apricot mixture and cook over low heat until thick and puffy, about 15 minutes. Remove from stove, stir in vanilla, crumbs, and nut mixture and coconut. Form into two rolls about 17cm (7″) long. Wrap in plastic or foil and store in refrigerator. To serve, slice as for Swiss rolls. Makes about 36 slices.

Dried Apple Pie

750g (3 cups) dried apples
625g (2½ cups) water
180g (¾ cup) sugar
1 tsp cinnamon
pinch nutmeg
pinch allspice 1 tbsp cornflour (cornstarch)
⅛ tsp salt
Pastry for 22cm (9″) 2-crust pie
1 tbsp lemon juice
2 tbsp butter

Reconstitute apples. Drain water and measure apples. (You will need 1500g (6 cups).) Combine remaining ingredients, then add apples and mix thoroughly. Spread in unbaked pie shell, sprinkle with lemon juice, add top crust, and seal edges. Cut air slits in top, and spread with 2 tbsp butter. Bake at 230°C (450°F) Mark 8 for 10 minutes, then lower to 173°C (350°F) Mark 4 and bake for 1 hour.

Pie Crust I

750g (3 cups) flour
250g (1 cup) solid shortening
4–5 tbsp ice water
½ tsp salt

Cut shortening into flour and add just enough water so you can handle dough easily. Divide dough in half. Roll out each half on floured surface to a thickness of 3mm (⅛″). Line pie plate with one of the rolled-out crusts, and prick with a fork. Fill, and cover with second crust. Press down around the edges to seal. Cut slits in top crust.

Pie Crust II

250g (1 cup) flour
⅛ tsp salt
6 tbsp butter or other solid shortening

Combine all ingredients with your fingers until the mixture resembles coarse sand. Press into 22cm (9″) pie plate, and prick with a fork. If a pre-baked crust is desired, bake for 10 minutes at 230°C (450°F), Mark 8.

Bananas

Dried bananas are very popular throughout Central America, and there is no reason why we 'gringos' should not also enjoy their amazingly sweet, concentrated flavour. As with all dried fruits they can be eaten just as they come out of the dryer. Use for flavouring puddings, ice cream, cakes and icings, breads, and milk shakes.

Blend dried bananas with just enough milk or water to cover blender blades, then add to recipe according to taste.

The following recipe makes a delicious light dessert.

Banana Whip

2 tbsp gelatine
310ml ($1\frac{1}{4}$ cups) milk
125ml ($\frac{1}{2}$ cup) hot water
500g (2 cups) blended banana mixture, as above
3 tbsp orange juice
3 tbsp honey
pinch of salt

Add milk, orange juice, honey and salt to banana mixture and blend thoroughly. Dissolve gelatine in hot water and stir into banana mixture. Pour into mould and set in refrigerator until firm. Serve with whipped cream sprinkled with nutmeg or cinnamon.

Lemon-Apple Fluff

Dried orange or lemon rind may be used just as you would the fresh product. Powder in blender, then add to frostings, batter, punch, anytime you need its sharp tantalizing flavour to spice up a dish or drink. Try this easy and surprisingly good dessert, using dried apples and citrus rind.

Reconstitute 180g (6oz) dried apples. Put through blender and add:
250g ($\frac{1}{2}$ lb) brown sugar
30g (1oz) butter
$\frac{1}{8}$ tsp nutmeg
1 tbsp dried lemon (or orange) peel

Blend all ingredients well, pour into saucepan and stir over medium heat for 3 minutes. Remove from fire and whip until very stiff. Drop by teaspoons on greased baking sheet. Bake in very low oven, 120°C (250°F) Mark 2, about 15–20 minutes. These featherlight confections will keep if stored in airtight containers.

Fruit Toppings

Blend any dried fruit with enough water to cover blender blades. Add a dash of lemon juice if the fruit is too sweet. After blending, add chopped nuts or sunflower seeds, sesame seeds or pumpkin seeds. Use as a sauce on ice cream, puddings, cake, pancakes and waffles. Also makes a nice surprise filling for omelettes.

Preserves

With dried fruit you can make up your jam as needed, instead of standing over a hot kitchen stove for hours during the summer months. You will also eliminate the need for large quantities of jars, lids and storage space. Make up several jars at a time and store in freezer or refrigerator. If sealed in sterilized jars, store on pantry shelf.

Soak dried fruit overnight. Next day, *simmer* in the same soaking water, adding more as needed, until fruit is fairly soft. Put through blender or chop in small pieces, depending on individual taste. Add sweetening as desired, also lemon juice if you wish. Pour in sterilized jars and cover either with paraffin, waxed or greaseproof paper or with a regular jar lid. Unless you use a tight seal, of course, this jam should be used before it has a chance to spoil.

Suggested combinations of fruit:

apple and pear; peach and pineapple; plum and apple (apple pie spices are good in this); plum and pear; apricot and pineapple.

Fruit Compotes

For a superb breakfast treat, soak any combination of chopped dried fruits overnight.

Try: prune, apple and raisins; apricots, peaches and pears; dates and apples, with chopped nuts or seeds sprinkled on top after the soaking period.

Vegetable Recipes

Greens

Dried vegetables like spinach, kale, broccoli, beet tops, swiss chard and cabbage should be simmered, never boiled, in just enough water to cover, until tender. Add more water if necessary during the cooking. Serve with butter and salt.

Meat and Vegetable Stew

(The following ingredients are suggestions only. When I make stew I throw in all kinds of leftovers. The vegetables can vary according to availability at the market and what I have in my drying larder.)

125g ($\frac{1}{2}$ cup) dried potatoes
80g ($\frac{1}{3}$ cup) dried carrots
80g ($\frac{1}{3}$ cup) dried tomatoes, sliced or quartered
1 tbsp (or more according to taste) dried celery
1 tbsp (at least!) dried onions
1500ml (6 cups) water
2 tbsp seasoned salt
1 tbsp bouillon liquid, or 4 bouillon cubes or any of the instant broth substitutes on the market
$\frac{1}{2}$kg (1lb) diced beef or lamb *or* TVP (textured vegetable protein, a meat substitute made from wheat gluten and soy beans. Measure TVP *after* it has been reconstituted, or follow directions on package)

Bring water to boil, add all ingredients and simmer 30 minutes or until done. I like stew a little better the second day—the flavours seem to have blended a little better by then, so why not make enough for two meals at one time? Or make a meat pie with what is left over.

Seasonings

Powdered dried vegetables make an excellent seasoning for almost any savoury dish. Make up your own combinations from your dried vegetables, starting with a base of tomato, parsley, onion, garlic and celery. Keep tightly covered on your kitchen shelf for instant added flavour. Also good for camping trips.

Powdered dried mushrooms, alone or in combination, are a real addition to the gourmet cooks' space shelf.

Instant Soup

Blend any desired combination of dried vegetables to a powder. To a cup of boiling water add 1 tbsp powder with any of the following: salt to taste; 1 tbsp nutritional food yeast; 1 tsp wheat germ; 1 tsp ground nuts or seeds. This makes a good between-meal snack for the dieter or for a convalescent patient. Strain if desired, as sometimes the vegetables, if not completely dry, will reconstitute and leave a thick residue at the bottom of the cup. Powdered mushrooms are very good added to this instant soup.

Blender Recipes

Only your imagination will set the limit to the combinations you can make with your dried foods. If you don't own a blender, consider the small food mills for sale in many health stores and the houseware sections of department stores. They do a good job of grinding dried food (it should be bone-dry if possible) to powder.

Combine one or more dried fruits with nuts, seeds or rolled oats or wheat germ for a delicious, healthful drink. Use water, juice or milk, and add honey or sugar if desired.

Use your blender to experiment with 'leather' variations. (Fruit leathers are puréed fruit, spread on plastic-lined dryer trays and dried, then rolled up and stored, usually in the freezer if over a long period of time.) Experiment by adding coarsely ground wheat to make crisp, chewy biscuits. Although you can add more sweetening according to taste, we suggest you try to keep the sugar amount low, not only to keep calories down but also so that family members who may have to restrict their sugar intake can also enjoy these delicious treats.

Sweet Corn

Dried sweet corn is reconstituted by soaking in water overnight, and tastes almost exactly like fresh corn just off the cob. It can be eaten just heated up with a little butter and seasoning added.

Corn souffles, corn chowders and corn puddings are just a few of the tasty, nutritious ways to use dried sweet corn. Use half the amount of dried corn in recipes calling for fresh corn, and reconstitute first.

Don't overlook the unbeatable flavour of freshly ground corn meal made from your own dried corn. If necessary, use your blender, but for best results invest in a small grinder which will reward you with delicious corn bread. In the UK it is sometimes easier to buy unpopped popcorn and grind it for the corn recipes below.

Spoon Bread

For 250g (1 cup) corn meal, grind 250g (1 cup) dried corn. Pour 375ml (1½ cups) boiling water over corn meal, then beat thoroughly and set aside to cool.

Measure out:

1 tbsp cooking oil, margarine or butter
1 tsp baking soda
1 scant tsp salt

Beat 2 eggs into corn meal mixture, then add other ingredients. Combine thoroughly, dissolving all lumps. Pour into a *preheated* greased baking dish and bake at 175°C (350°F) Mark 4, for 30–35 minutes. Centre is completely done when a silver knife inserted comes out dry.

Serve with a large salad and plenty of fresh butter for a splendid luncheon dish.

Granola

Heat:

125ml (½ cup) honey
125ml (½ cup) oil
2 tbsp milk

Do not boil. Pour over the following well-mixed ingredients.

250g (1 cup) coconut (freshly grated if possible)
125g (½ cup) untoasted wheat germ (Always keep untoasted wheat germ refrigerated until used, to avoid rancidity)
30g (¼ cup) sunflower seeds
30g (¼ cup) sesame seeds
30g (¼ cup) pumpkin seeds
250g (1 cup) chopped walnuts *or* substitute almonds, filberts or raw cashews
1kg (4 cups) rolled oats (not the quick-cooking kind)

Mix thoroughly until all dry ingredients are coated with the heated mixture. Spread on dryer trays which have been lined with plastic wrap, greaseproof paper or waxed paper. Granola is done when it is crumbly and can be easily separated. If you are in a hurry, fill only one half of the trays, staggering them so alternate sides of the trays are empty to allow for better air circulation. Cool before storing.

Homemade Noodles

2 eggs slightly beaten with
scant ½ tsp salt
Add to:
330g (1⅓ cups) flour

Beat into a fairly stiff dough. Knead a few times, then let stand in covered bowl 45 minutes. Roll out into very thin sheets, let sit until the dough is a little dry but not hard. Cut into strips according to your own preference as to width. Thin strips will cook faster. Spread on trays, without overlapping the pieces, and dry until very brittle, a little dryer than commercially prepared noodles.

These noodles will cook up quite nicely. Store in airtight containers and cook as for bought noodles.

Variations: Try a combination of 4 parts wheat flour to 1 part soy flour for added protein and chewiness. If you can find a place to buy Jerusalem artichoke flour, add to wheat flour in same ratio for a noodle lower in starch.

For a really high protein noodle, use 250g (8oz) gluten flour, 500g (16oz) soy flour, 6 eggs, and 1½ tsp salt. Use whole wheat flour (it will take a surprisingly large amount) for rolling out dough and cutting strips.

Steamed Sponge Pudding with Fruit or Purée Layers

125g (4oz) margarine
125g (4oz) sugar
250g (8oz) self-raising flour
2 standard eggs
125g (4oz) prepared dried fruit, puréed or not as you wish

Cream fat and sugar until pale, light-coloured and fluffy. Beat in eggs one at a time. Fold in flour. If necessary add 1 or 2 tbsp milk until batter is a soft, dropping consistency.

Spread one-third of the mixture into the bottom of a greased 800ml (1½pt) basin. Cover with half of the fruit or purée, cover this with half of the remaining cake mix, then the rest of the fruit, then the last of the cake. Cover with foil, place in saucepan of boiling water (or steamer) for 1½ hours. Alternatively, place in a baking tin and bake in the oven at Mark 3–4 (350°F) for 35–40 minutes.

Variation: Using only half the quantity of fruit, place it in the bottom of the bowl or tin and put all the cake mix on top. When baked the cake turns out with a fruit 'cap'. Serve either variation with custard or cream.

Fruit Crumble

250g (8oz) dried fruit (reconstituted and chopped)
1 tsp grated lemon rind (¼ tsp dried rind)
125g (4oz) flour
60g (2oz) fat
60g (2oz) sugar
1 tsp cinnamon *or* 30g (1oz) dried coconut

Place flour, fat, sugar and either cinnamon or coconut in a bowl. Rub together with fingertips until they resemble breadcrumbs.

Combine fruit and lemon rind, and spread over bottom of greased bowl. Pile crumble mixture on top. Bake at 190°C (375°F) Mark 5, for approximately 30 minutes.

Fruit Capped Suet Pudding

125g (4oz) suet (or other shortening)
60–80g (2–3oz) sugar
250g (8oz) self-raising flour
milk
dried fruit: 125g (4oz) reconstituted, or 25–60g (1–2oz) dried, then reconstituted and puréed if desired

Place fruit or purée in the bottom of a greased 500ml (1pt) bowl. Place suet, flour and sugar in a bowl and mix with enough milk to make a dropping consistency. Turn into the bowl on top of the fruit, cover with foil and steam 1½–2½ hours until light and spongy.

Savoury Suet Pudding

60g (2oz) suet
125g (4oz) self-raising flour

375g ($\frac{3}{4}$lb) chopped meat or 125g (4oz) dried meat reconstituted
or 125g (4oz) TVP (soya protein granules)
2 tbsp dried onions
seasoning
stock or bouillon cube as required

Make pastry and line bowl as for fruit-filled pudding. (See previous recipe.) Combine meat, seasoning and stock. Fill the pastry-lined bowl with the meat mixture. Seal with pastry top, then make small hole for steam to escape. Cover with foil and steam $1\frac{1}{2}$–$2\frac{1}{2}$ hours or until meat is tender.

Note: Meat can be sautéed with onion first to shorten cooking time.

Dried Apple-Oatmeal Drops

375g ($1\frac{1}{2}$ cups) dried apples (reconstituted)
250g (1 cup) raisins combined with chopped nuts or sunflower seeds
125g ($\frac{1}{2}$ cup) butter or margarine
250g (1 cup) dark brown sugar
2 eggs
430g ($1\frac{3}{4}$ cups) flour
125g ($\frac{1}{2}$ cup) rolled oats
$\frac{1}{2}$ tsp salt
2 tsp baking powder
$\frac{1}{2}$ tsp cinnamon
$\frac{1}{2}$ tsp allspice

Cream butter and sugar, add eggs and beat thoroughly. Mix all other ingredients in separate bowl, then add to first mixture. Drop by spoonfuls on oiled baking sheet and place in preheated moderate oven 175°C (350°F) Mark 4, until done, 11–14 minutes. Yield: 3–4 dozen. This is another exceptionally popular recipe with all ages.

Graham Crackers or Digestive Biscuits

1kg (4 cups) whole wheat flour
250g (1 cup) butter
250g (1 cup) dark brown sugar

1 tsp baking soda
1 tsp cream of tartar
1 lightly beaten egg
125ml (½ cup) hot water

Cut butter into flour until mixture is like coarse meal. Add all other ingredients, but only enough water to make dough into pastry consistency. Roll out on a floured surface till very thin (about 3mm (⅛″) to 6mm (¼″) thick). Cut into small squares, depending on what size you prefer. Bake on *unoiled* baking sheets until golden brown, about 15 minutes. Cool on rack. Yield: 2 dozen. These crackers are very popular in the US for babies, but I have never seen them in the UK. They are delicious spread with cream cheese and apricot preserves—or strawberry, or redcurrant jelly, etc.

Special Tasty Spice Cake

250g (1 cup) dried apples (dates, figs or prunes can also be used)
1 tsp cinnamon
1 tsp ginger
1 tsp cloves or allspice
180ml (¾ cup) water
180g (¾ cup) butter or oil
375g (1½ cups) brown sugar
3 eggs
750g (3 cups) flour
5 tsp baking powder
½ tsp salt
250ml (1 cup) fresh milk
1 tsp vanilla essence (extract)

Wash fruit, chop fine. Add spices and water and boil for 3 minutes, stirring constantly. Cool. Cream shortening and sugar with light and fluffy. Add one egg at a time, beating well. Sift remainder of dry ingredients together, then add alternately with milk to the creamed mixture. Do not overbeat, only until smooth.

Fold in fruit and vanilla essence (extract), place in 3 buttered 22cm (9″) sandwich (cake) tins and bake at 190°C (350°F) Mark 4, for 25 minutes, until done. Serve with puréed dried fruit, such as apricots or peaches sweetened if necessary with a bit of honey, and with whipped cream on top. Also very good with Peach Cream Filling.

Peach Cream Filling

(or apricot or pear or pineapple or apple)

180g (¾ cup) reconstituted dried peaches (measure after soaking)
6 tbsp brown sugar
3 tbsp cornflour or 1½ tbsp arrowroot
⅛ tsp salt
½ tsp vanilla flavouring
1 large egg
250ml (1 cup) scalded milk
2 tbsp butter
125g (½ cup) chopped nuts (optional)

Combine all dry ingredients and then mix with slightly beaten egg. Stir in just enough hot milk to make a thin paste. Add to the rest of the hot milk and cook over boiling water 5 minutes, stirring constantly. Add butter, cool and fold in finely chopped fruit. Now cook for another 10 minutes (it may not take this long) until mixture is thick, stirring constantly. Add vanilla, and use as a filling for cream puffs, cakes or topping for the above spice cake.

Purée of Apricots

250g (1 cup) dried apricots
1 tbsp honey
1 tsp grated orange rind (dry or fresh)

Cook fruit until tender in small amount of water to preserve flavour. Liquidize, add honey and grated orange rind and mix thoroughly. Chill. Serve plain, with cream, or as topping for puddings, cakes and custards.

Fruit Nut Bars

500g (2 cups) chopped dried fruit
4 eggs, well beaten
330g (1⅓ cups) honey
330g (1⅓ tsp) grated lemon rind (dried or fresh)

330g ($1\frac{1}{3}$ cups) pastry flour
$\frac{1}{3}$ tsp salt
$\frac{1}{3}$ tsp baking powder
125g ($\frac{1}{2}$ cup) nuts (walnuts, pecans, hazelnuts)

Add honey and lemon rind, finely grated or crumbled, to well-beaten eggs. Beat again. Sift dry ingredients and add to egg mixture. Fold in chopped fruit and nuts and pour into 20 × 30cm (8″ × 12″) pan that has been greased, lined with greaseproof (waxed) paper, then greased again. Bake at 175°C (325°F) Mark 3, about 45 minutes. Turn out to cool on rack. Remove paper. When cool cut into bars and frost if desired. Makes about 3 dozen. A good recipe for lunch baskets, picnics or travelling.

Apricot Snow

500g (1lb) dried apricots
250g (1 cup) honey
3 stiffly beaten egg whites

Purée soaked apricots and honey. Fold in egg whites. Pour into greased mould or refrigerator tray and freeze. Serve frozen, but break up into bite-size pieces. Garnish with whipped cream and/or sliced bananas.

Easy Prune Whip

A lovely light dessert, good for children on a hot day, and very easy to make.

250g (1 cup) prunes
60g ($\frac{1}{4}$ cup) sugar or 3 tbsp honey
$\frac{3}{4}$ tsp vanilla flavouring
60g ($\frac{1}{4}$ cup) chopped nuts (optional)
3 egg whites
Cream, plain or whipped

Soak prunes overnight, or simmer until soft. Remove stones, mash or put through blender or food mill, then simmer again, with sweetening added, for 15 minutes. While prunes cool beat egg whites until stiff, then fold into prunes and add vanilla, chopped nuts, and cream on top. Also good with a dash of grated orange peel sprinkled on top of the cream.

Apricot Cream Whip (also good in peaches)

Another delicious dessert, but not recommended for low-calorie diets.

500g (1lb) dried apricots
125g (½ cup) sugar or 4 tbsp honey (or to taste)
1 tbsp lemon juice
250ml (1 cup) cream (evaporated milk may be substituted)

Pour boiling water over apricots and simmer in just enough water to keep them from burning, until soft. Liquidize, add sweetening and lemon juice. Bring to boil to be sure all ingredients are thoroughly heated, cool, then fold in very stiff whipped cream.

Fruit Sherbet

285ml (½pt) fruit purée
2 tsp lemon juice
1 egg white
sweetening if desired

Stir lemon juice with purée. Place in tray in deep freeze or freezer compartment of refrigerator. When purée is frozen 12mm (½″) around the edge of pan, remove and stir until the frozen part is broken up and a bit mushy.

Whisk egg white until stiff. Add to purée and mix evenly. Pour back into freezer tray and refreeze until firm.

Apricot Fool

125g (4oz) dried apricots
1 tsp lemon juice
3–4 tbsp water
30g (1oz) sugar
115ml (¼pt) thick custard, sweetened
115ml (¼pt) double cream (whipping cream)

Simmer apricots in water until soft. Add lemon juice and sugar. Purée fruit and let cool. Add custard and mix. Whisk the cream and fold into fruit mixture, but save out some for decoration, if desired. Serve in small individual bowls, or use as centre piece for a buffet dinner.

Sunshine Peach Pudding

250g (1 cup) dried peaches (or other dried fruit)
125g (4oz) chopped nuts
60g (2oz) cornflour (cornstarch)
2 large eggs, well beaten
2 tsp sugar
125ml (½ cup) cream
125ml (½ cup) water for soaking peaches

Chop soaked fruit finely. Combine cornflour, sugar, and nuts. Combine liquids, add eggs, then add all but the fruit, and mix. Stir in fruit, place in buttered baking dish and steam about an hour, or until done. Serve with cream or fruit sauce. Good with ice cream, too.

Dried Fruit Custard

125g (4oz) dried apples (reconstituted)
8 dried figs (soak overnight)
3 large eggs
375ml (1½ cups) milk
2 tbsp honey
1 or 2 whole cloves
½ tsp cinnamon
dash of nutmeg if desired

Place milk and honey in top of double saucepan and while it warms, beat the eggs, then add to milk. Cook, stirring constantly, until thick, then stir in finely chopped apples. (You can liquidize them, but the custard tastes better with bits of apple floating around.) Add spices and stir. Place sliced figs on the bottom of a flat dish, such as a large glass pie plate. Pour custard mixture over figs, let set until firm. Garnish with whipped cream, fresh orange slices, chopped nuts, etc.

Uncooked Christmas Pudding

2 tbsp of *each* of the following, all very finely chopped:
figs
dried apples
dried cherries

raisins
sultanas (white raisins)
dates
apricots
pears
2 tbsp lemon juice
(If some of the above are not available, substitute what you do have in equal amounts, but try to have a variety.)
250g (½lb) fresh green seedless grapes
125g (4oz) walnuts, almonds *or* pecans (hazel nuts also if available are good in this, but you can also combine any or all of these nuts)
2 very ripe fresh bananas or the equivalent of dried reconstituted bananas. Estimate amount of dried fruit needed as this doesn't need to be all that exact
750g (1½lb) muesli, *or* Granola mixture *or* bran and raisin mixture
1 tsp ginger
1 tsp nutmeg

Combine dried fruit, add chopped nuts and grapes. Mash bananas, add to fruit mixture, then add spices and lemon juice.

Crush grain mixture by rolling between two sheets of grease-proof (waxed) paper with jar or rolling pin, then add to other ingredients and mix really well. Pack into a buttered baking dish, press down well, cover and let it stand overnight in cool place.

Topping:
2 bananas (if dried, reconstitute and then mash)
750ml (3 cups) thick cream, stiffly whipped
grated nutmeg to taste
1 tbsp lemon juice
1 egg white, stiffly beaten

Combine mashed bananas with whipped cream, then fold in egg white to which lemon juice and nutmeg have been added. Mix all well but very lightly. Garnish pudding.

Steamed Fruit Pudding

250g (8oz) of *each* of the following:
dates

raisins
sultanas (white raisins)
dried apples
125g (½ cup) dried bananas
1 tsp dried orange peel (or oz fresh)
30g (1oz) butter, melted
2 large eggs, well beaten

Grind or chop dried fruit very fine. Add butter and eggs, mix thoroughly. If mixture appears a bit stiff, add a dash of milk or fruit juice. Place in buttered baking dish, cover with foil, place in basin of hot water in oven. Steam until thoroughly cooked, at least 2½ hours. Serve either hot or cold, with cream or a fruit sauce.

Creamed Pear Delight

500g (1lb) dried pears
1 tbsp lemon juice
3 egg whites, beaten very stiff with ¼ tsp cream of tartar
2 tbsp sugar
2 tsp cinnamon
dash of salt
½ tsp vanilla essence (extract)

Simmer pears until soft and liquidize. Add lemon juice and sweetening, stir well, and then gradually add to stiffly beaten egg whites. Chill. Serve with whipped cream and finely chopped nuts.

Hors D'Oeuvres Prunes

Soak the largest prunes you can find overnight. The next day remove the stones and stuff with the following mixture:

sour cream or natural yogurt
finely chopped apple (if using dried apples, reconstitute)
chopped nuts such as pistachio, almond, walnut or hazel nuts

Serve as a surprisingly delicious and different party appetizer.

Tarts with Dried Fruit Filling

60g (2oz) dried apples
60g (2oz) dried pears
60g (2oz) dried apricots

60g (2oz) dates (stones removed)
2-crust pie dough recipe

Pour boiling water over all fruit except dates, and simmer for 20 to 30 minutes. Set aside to cool, add dates, then chop in blender but *do not purée*. Add sweetening if desired, although usually the dates are sufficient.

Make up pie crust, divide into small amounts, and roll thinly on lightly floured board. Cut into 12½cm (5″) squares, place 2 tbsp filling on one corner, then fold over to make a triangle. Pinch edges to seal. Dab edges first with a bit of cold water if necessary to make a tight seal. Place on baking sheets and bake in 175°C (350°F) Mark 4 oven until done, 30–35 minutes. Very good for lunch baskets, or to feed a horde of hungry Scouts. Our family loves these and you really can't ever make too many.

Birrenbrot (Bavarian Dried Pear Bread)

250g (8oz) dried plums, pears, figs, raisins, sultanas (white raisins), dates (stones removed), (total weight for all fruit)
100g (3½oz) hazel nuts
100g (3½oz) walnuts (substitute hazelnuts if necessary)
100g (3½oz) almonds
75g (2¾oz) lemon or orange rind, chopped (fresh or dried)
1 tsp cinnamon
1 tsp cloves
1 tsp aniseed
5 tbsp rum
sugar to taste
500g (1lb) black bread dough
(recipe on page 141)
(If not desired or unavailable, plums and/or dates may be omitted)

Wash plums, pears and figs. Soak overnight and cook the next day until *half* soft.

Slice plums and pears. Wash raisins and sultanas, and cut all fruit into large pieces. Blanch almonds. (To blanch, pour boiling water over shelled almonds and slip off the skins.) Halve walnuts, but leave the hazelnuts whole.

Mix all ingredients with black bread dough. Knead very well.

Our Bavarian friend who gave us this recipe told us, 'Knead until you have lost one kilo!' When ready, set aside one-third of the dough. Form remainder into a round loaf and place on a flat baking sheet that has been greased and lightly floured. Let rise until double in bulk. Now roll out the remainder of the dough until it is *very thin*. Brush fruit water on top, and place dough on top of the formed loaf. Prick holes with fork. Bake at 175°C (325°F) Mark 3 for 1½ hours. Immediately after baking brush bread with *hot* fruit washings. (Keep the water the fruit was soaked in for this.) This recipe makes nice Christmas presents.

Black Bread

- 2 packets dry yeast or 60g (2oz) fresh yeast
- 125ml (½ cup) warm water
- 2 tbsp sugar or honey
- 4–6 tbsp treacle (dark molasses)
- 500ml (2 cups) hot water
- 375g (12oz) rye flour
- 180g (6oz) whole wheat flour
- 45g (1½oz) soy flour can be substituted for an equal amount of wheat flour for added protein
- 5 tbsp oil or melted butter
- 2 tsp salt

Step 1: Dissolve yeast in the *warm* water. (Hot water will kill the yeast and the bread will not rise.) Add sweetening. Stir until thoroughly dissolved and set aside for 10 minutes. At the end of this time, the mixture should be frothy and bubbling. If it is not, the yeast is not active and you must mix up more, hoping the next yeast will be alive. Otherwise the bread will be flat and hard.

Step 2: Dissolve treacle (molasses) in hot water, and to this mixture add flour, shortening and salt.

Step 3: When flour and treacle (molasses) combination is well mixed, add yeast and mix again very thoroughly.

Step 4: Turn dough out onto floured surface and knead vigorously, adding small amounts of flour as needed to keep dough from sticking. You will probably be surprised at how much extra flour this step will take. Dough should be springy, elastic and pliable after about 15 minutes. Place in greased bowl, cover with

towel and set in warm place to rise. In the UK, the airing cupboard is a good place for this. You can also use your food dryer, but watch the thermometer as bread should not rise in any setting over 25°C (80°F). When dough has doubled in bulk, punch down and let rise again. Form into loaves and bake at 145°C (325°F) Mark 3 for 1½ hours or until done.
Note: If you want to make just the bread, and would like a good additional German touch, add 60g (¼ cup) caraway seeds to the dough before kneading. For a shiny finish, brush surface before baking with either melted butter or beaten egg yolk.

Celery Soup

60–125g (2–4oz) dried celery
30g (1oz) dried onions
285ml (½pt) water
1 bouillon or soup cube
Substitute vegetable or meat stock if desired
Salt and pepper to taste

Reconstitute dried vegetables first, *or* simmer gently in the liquid until softened. Add seasoning.

For a very smooth soup, put in blender when finished. Serve with croutons or float a small heap of grated cheese on top of each individual serving.

Mushroom Soup

(Mushrooms are one vegetable of which you can never have too much. Dry them at all times of the year, whenever you find them at a bargain price.)

1l (1 quart) stock (use soaking water from mushrooms if desired)
125g (¼lb) dried mushrooms (reconstituted until soft)
30g (1oz) butter or oil
250ml (1 cup) natural yogurt or sour cream
1 large egg
1¼ tbsp flour
Salt and paprika to taste
Chopped parsley if desired (about 2 tbsp)

Cook mushrooms in heated butter with paprika and salt until

tender. Sprinkle the flour over them and add the parsley, then cook 1 minute. Add stock, stir constantly until smooth. Add 2 tbsp yogurt or sour cream, simmer 45 minutes. Mix the rest of the yogurt or sour cream with the beaten egg in a large bowl, then slowly pour the still boiling soup over the mixture, stirring constantly. (You may need some help with this last bit, as the recipe comes from a very old cookbook, written when the cook had plenty of kitchen helpers.)

Sprinkle with more chopped parsley and serve immediately.

Tomato, Onion and Noodle Soup

1l (1 quart) stock
60g (2oz) butter or oil
3 fresh tomatoes or 125g (½ cup) dried tomatoes
3 fresh onions or 125g (½ cup) dried onions
Salt, pepper and paprika to taste
250g (1 cup) dried noodles, boiled until tender

Chop onions, then fry in butter until golden brown. Stir in seasonings, add stock and tomatoes and simmer until tender and tasty, about 1 hour. Tomatoes should be first softened before adding to mixture.

You can either strain the soup and serve clear, or eat everything which is what we like to do. Add noodles, already heated, just before serving.

Cold Finnish Fruit Soup

250g (1 cup) dried apricots
250g (1 cup) prunes
250g (1 cup) raisins
80g (⅓ cup) honey
1 tbsp gelatine or 1 tbsp agar-agar
4½l (1 gallon) water

Chop fruit and cook until soft. When nearly cooked add honey. Soften gelatine or agar-agar in cold water and add to soup. Stir thoroughly. Let cool, beat well and serve quite cold. Serves 6. A nice appetizer for lunch on a warm day.

Potato and Onion Soup

125g (4oz) dried potatoes
125g (4oz) dried onions
Salt and pepper to taste

'Bouquet garni' made by tying into a small cloth bag such herbs as rosemary, marjoram, thyme and basil. Bring 2l (2 quarts) water to boil. Toss in dried vegetables, seasonings and herbs and cook over low heat until tender. If you like a very thick soup, add some flour mixed with cold water to make a smooth paste, to the soup. When finished, blend for a delicious smooth thick soup, or eat it as it is. Delicious served with crusty rye bread.

An interesting suggestion from a Dorset housewife on thickeners: to make an extra supply of 'beurre manié' combine 2 parts softened butter to 1 part flour, and mix thoroughly. Store in refrigerator and when you need a thickener, as for the above soup, just stir in until smoothly dissolved.

Beetroot Bortsch

If you grow your own vegetables, why not plant extra beetroot and enjoy lovely Russian-style bortsch all year round, with your own home-dried beets?

When I asked a friend from 'the old country' for a recipe for real Russian bortsch, I received the answer of a true cook—just enough of everything to make it taste good. I never could pin her down, for example, as to how much water or stock to start with. All she said was, 'Make plenty, enough for at least two meals'.

The best advice is to keep tasting as the soup cooks, and if it seems a bit strong add more water. Start with a good stock, using a beef bone or save in advance any water you have cooked vegetables in. Flavour with seasoned salt or any of the good soup mix bases on the market.

Reconstitute and dice any or all of the following vegetables: onions, carrots, leeks.

250g (1 cup) dried beetroot (1 or 2 very large fresh ones)
(You can use more of this if you wish, after all, it is called beetroot bortsch)

Add the diced vegetables to 4l (1 gallon) of stock, plus 2 tbsp apple cider vinegar or lemon juice. Simmer until soft, adding salt, pepper (cayenne is very good in this) and just before serving add some grated *raw* beetroot if it is the right time of year for it.

If you serve this soup hot, add the grated pulp just before serving, bring to a quick boil to warm the pulp.

It is really better served icy cold with large dollops of yogurt or sour cream floating on top.

Apple Soup (Old German Recipe)

375g (12oz) dried apples, reconstituted
2 thick slices stale bread
1l (1 quart) water (include soaking water from apples)
125g (½ cup) raisins
3 tbsp butter
salt and sugar to taste
2 whole cloves
1 tsp cinnamon

In large cooking pan add bread, apples, water, spices and salt. Cook very slowly until apples are quite soft. Remove from pot, liquidize, and return to pot, adding butter, raisins and a dash of sugar. Bring to the boil, then simmer until raisins are plump. If desired, sprinkle in a few almonds. Serve with dark bread fried in butter.

Egg and Vegetable Platter

250g (1 cup) dried mushrooms, reconstituted 250g (½lb) fresh
4 tbsp dried onions, reconstituted (2 small fresh)
4 large eggs
60g (2oz) grated cheese
375g (1½ cups) dried mixed vegetables (625g (2½cups) fresh)
(Use one of your soup mix packets)
375ml (1½ cups) gravy or white sauce
1–2 tsp mixed herbs

Simmer vegetables, with salt and pepper to taste, until tender. A few minutes before they are done, add the herbs and stir well. While they are finishing, sauté first the onions, then add the mushrooms, stirring occasionally. Add well-beaten seasoned eggs stirring gently until cooked but still soft.

To serve: Arrange cooked vegetables around the edge of a platter, and place scrambled egg mixture in the centre. Pour sauce over vegetables. Serve grated cheese in separate bowl.
Note: Any combination of fresh and dried vegetables can be used for this dish Serves 4.

Two Recipes from Sri Lanka

Chillies in Clear Yogurt Use long, thin green chillies. In the morning parboil in salted water to cover, to preserve the colour. That same evening, immerse the chillies in a bowl of yogurt. Next morning remove from yogurt and place in dryer. Return to the same yogurt every evening for three days, placing in dryer again the following morning. At the end of three days throw away the yogurt and eat the chillies—delicious!

Dried Potatoes with Sweet (Bell) Peppers Reconstitute dried potatoes, then mash with butter and salt and a little milk. Remove fibres and seeds from inside the sweet (bell) peppers, and stuff with mashed potato mixture. Dip peppers in batter and deep fry. Makes an unusual and very tasty side dish.

Curries

Delicious recipes for curry dishes are almost unlimited with dried foods. We found this recipe a bit out of the ordinary.

Tomato, Egg and Bean Curry

250g (8oz) beans—lima, broad, haricot or navy beans will do
Dried green beans are also good. Reconstitute enough to make 500g (2 cups)
4 fresh tomatoes or enough dried tomatoes or leather to reconstitute into about 185g (¾ cup)
4 tbsp dried onions (1 large) fresh
125g (½ cup) dried apple, reconstituted, then chopped
2 tbsp butter or oil
2 tbsp flour
1 to 3 tsp curry powder (depending on how hot you like it!)
3 hard-boiled (hard cooked) eggs, chopped in large pieces
500ml (2 cups) water or stock
1 tbsp lemon juice

Soak beans overnight. Put on to cook in plenty of water, and while they cook, reconstitute onion in water, then fry in the butter or oil until golden. Add curry powder, flour and chopped apple. Cook all together for 5 minutes. Add liquid, then the drained beans, egg, tomatoes and seasoning. Simmer until all ingredients are thoroughly heated. Before serving, stir in the lemon juice. Serve with rice and little side dishes of chutney, peanuts, raisins and yogurt with grated cucumber and mint leaves. Bananas with lemon juice squeezed over them can also be served as a side dish.

Basic Vegetable Casserole

125g (4oz) dried potatoes 1kg (2lb mashed)
125g (4oz) dried carrots 250g (½lb raw, grated or shredded)
4 tbsp onions dried (2 medium fresh)
250g (1 cup) dried green beans 250g (½lb fresh)
Note: These are only suggested amounts. You can use any ratio you wish of your favourite vegetable
90g (3oz) soy flour (instead of regular flour, because of its high protein content)
155ml (5fl oz) milk
½ tsp salt (or to taste)
1 tbsp mixed herbs
500g (1lb) mince (hamburger) or chopped leftover meat
Bouillon cubes, Marmite or any prepared soup mixtures, to taste

Reconstitute and mash the potatoes. Place all ingredients except meat, soy flour and potatoes in saucepan, cover with boiling water and simmer until *half done*. Add more seasoning at this time if desired. Add meat. Pour into large glass baking dish. Make a smooth paste by combining soy flour and milk over very low heat. Mix this well with the mashed potatoes and spread over vegetables. Bake in moderate oven 175°C (350°F) Mark 4, until lightly browned, about 30 minutes. If desired, lightly fry some additional onions until crisp and brown and sprinkle on top just before serving, or sprinkle grated cheese on top and cook until cheese is melted and golden brown. Instead of meat you could lightly brown reconstituted TVP (soy protein granules) and combine it with potatoes after mashing.

Bibliography

Adams, Catherine, F. *Nutritive Value of American Foods in Common Units*, Agriculture Handbook No 456. US Department of Agriculture, Agricultural Research Service

Carrigher, Sally, *Wild Heritage*, Panther ed. 1967, Hamish Hamilton Ltd

Dickey, Esther, *Passport to Survival*, Bookcraft Publishers, 1848 2300 South, Salt Lake City, Utah, USA

Grange, Cyril, *The Complete Book of Home Food Preservation*, Cassell and Co. Ltd, London 1947

McCance, R. A. and Widdowson, E. M., Medical Research Council SRS 297, *The Composition of Foods*, HMSO

Mellanby, Kenneth, *Can Britain Feed Itself?* Merlin Press, 1975

Ministry of Agriculture, *Manual of Nutrition*, HMSO

Ministry of Agriculture, Fisheries, and Food, *Home Preservation of Fruit and Vegetables*, Bulletin No 21, HMSO

Oregon State University, *Drying Fruits and Vegetables*, Extension Circular 889, June 1976. Oregon State University Extension Service, Corvallis, Oregon, USA

Sholto Douglas, J. and Hart, R. A. de J., *Forest Farming*, Robinson and Watkins Books Ltd, 1976

Soil Association Quarterly, September 1975

Utah State University, *Home Drying of Fruits and Vegetables*, Extension Service, Utah State University, Logan, Utah, USA

Watt, Bernice K. and Merrill, Annabel L., *Composition of Foods*, Agriculture Handbook No 8. US Department of Agriculture, Agricultural Research Service.

Acknowledgements

My grateful thanks to the following friends and fellow food-drying enthusiasts who so generously shared their knowledge, experience and ideas with me.

Harriet and Warren Clingman, who started us drying food; Evan and Melba Garner; Ken and Shirlie Altman; Clint Seibert; Dick and Izetta Nesbitt; Bernie and Mary Masters; Mrs Sven Barton; Asher and Olive Lester; Alois Kennerknecht; John Bond and Janet Parry; our friends from Sri Lanka and Iceland and the editors at David & Charles for their assistance and encouragement. And to my family, without whom the book would certainly never have been written.

Drawings by Michael Sperinck.
Refrigerator drawing by Jacek Slezynski.
Photographs by Donald Nichols, David Gorvin and Tony Griffiths.
Small box dryer designed by Asher Lester and Donald Nichols.
Large box dryer reproduced by permission of Oregon State Extension Service, Corvallis, Oregon, USA.

Index